WORKSHOP PRO

# Setting up a home car workshop

# www.veloce.co.uk

First published in May 2018 by Veloce Publishing Limited, Veloce House, Parkway Farm Business Park, Middle Farm Way, Poundbury, Dorchester DT1 3AR, England. Tel +44 (0)1305 260068 / Fax 01305 250479 / e-mail info@veloce.co.uk / web www.veloce.co.uk or www.velocebooks.com.
ISBN: 978-1-787112-08-7; UPC: 6-36847-01208-3

# Setting up a home car workshop

JULIAN EDGAR

VELOCE PUBLISHING
THE PUBLISHER OF FINE AUTOMOTIVE BOOKS

# Introduction

I've been playing in home workshops all my life, and a lot of that time has been spent working on my own cars.

In addition to normal maintenance tasks, I've performed major overhauls and have spent a lot of time making car modifications. Often those modifications have required welding, machining and general fabrication, and so over the years I've acquired the tools and equipment that allow me do those things.

I've also moved house frequently, and so I have been in the (enviable?) position of building two new home workshops from scratch, and setting up another three. In this book, I cover some of the layouts and equipment of my various workshops.

I've also been helped by four people who have allowed me to use details of their home workshops – thanks to Jesse Holman, Russell Ferguson, Bill Sherwood and Jack Olsen. (My thanks also to Sam Sharp and Bill Sherwood who read and commented on the draft chapters.)

In researching this book, I came across a lot of home workshops and garages where work on cars doesn't actually happen! You could say in that time-honoured phrase that these workshops are 'for display purposes only' (perhaps, to be fair, they are really only for the storage of cars). This book is different. The workshop tools, equipment and layouts that you will see in these pages are designed to allow you to successfully work on your car. The storage, lighting and ventilation suggestions are based on what I have found to be effective, not what just looks good. These workshops may not be the sort where you can eat your dinner off the workbench, but they are also workshops where people – normal people on normal budgets and with normal skills – have achieved great outcomes.

I've also devoted whole chapters to aspects I think are critically important. Without a solid,

strong and immoveable workbench, you'll find it hard to fabricate anything. Welding is so vital that it gets its own chapter. You can use either ramps, hoists or pits to allow you to work underneath your car – and I have covered all three approaches. Any car of the last 40 years is likely to contain electronic as well as electrical systems, so I have a chapter on buying and using tools suitable for working on these vehicles as well.

I've also included a full chapter on safety – people die or are badly injured every year working on their cars at home, and I don't want you to be one of them. There's a chapter with some hints and tips – and another chapter that might puzzle you. It's called 'Approaches to Innovative Design' and covers some of the major hindrances I see to people achieving great outcomes in their home car workshops. I'll let you read the chapter for yourself to discover what those hindrances are.

I think that working on your car in a home workshop is one of the great pleasures of life. A home workshop is a place where you can lose yourself in a project, working with your hands and brain to achieve something unique and rewarding. Sometimes I become so engrossed that I am startled when I walk out of the workshop to find that life outside is still going on normally!

No matter your budget, I hope that you find many things in this book that help you develop that special and effective space.

---

**Footnote:** I live in a metric country and so my thinking in all measurements is metric. However, I understand how not being able to immediately picture size or weight can easily slow understanding, so I've provided Imperial conversions. I've chosen not to use mathematically 'pure' conversions, though; rather, where it makes better sense, to use approximations. This prevents strange sentences where the conversion appears to have decimal-places of importance that is misleading.

Julian Edgar

**Chapter 1**

# Buying and using hand tools

People who either have formal training in the use of hand tools (eg they've done a trade apprenticeship) or who have had long experience of using tools (ie they've learnt from their mistakes) tend to assume that everyone already knows about tools. I mean, what is there to know about using spanners and the like? Well, quite a lot.

It's only when you see people who have simply no idea of how to use hand tools that you realise the very real necessity of learning. I remember when I taught automotive mechanics to secondary school students. Not having a 'trade' background, I wondered if my ignorance would soon show, but I need not have worried. After all, it was only a few lessons into the term that I saw a student using an adjustable shifter (in the US, a crescent wrench) to try to remove an engine's head bolts … (and for those that wonder at the significance: head bolts are among the most tightly-torqued of any bolts in a car … you need a well-fitting socket and a long lever to undo them.)

Knowing how to use hand tools has two major advantages:

* You're more likely to achieve success
* You're less likely to damage the equipment you're working on.

And those two points are related: round-off the head of a nut you're trying to undo with the wrong tool and/or technique, and you'll find success terribly elusive …

In this first chapter, I want to look at some of the practical aspects of buying and using hand tools.

## A BASIC TOOL KIT

Most of you reading this will already have a wide variety of hand tools – screwdrivers, socket set, spanners, a hammer, and so on. But what if you've *not* got these tools – and you've got the funds to buy what you need?

The major tool kits sold by automotive suppliers give you a good guide as to what to buy. In this example, I'll assume that you are primarily working on a car equipped with metric fasteners, so you would not buy a combination metric/Imperial set, just the one with metric tools. (If you work primarily on cars with Imperial fasteners, make the obvious change, and if you work on both, expect to spend more money!)

The following listing shows a good, general-purpose automotive tool kit. It is based on the SP Tools '138pc Metric Custom Series Tool Kit.'

## SOCKETS

### ¼ inch drive sockets & accessories

* 12-point – 4, 5, 6, 7, 8, 9, 10, 11, 12 & 13mm
* 45 tooth ratchet & flex handle
* Wobble extension bars – 50 & 150mm
* Torx bits – T10, T15, T20 & T25
* Hex bits – 3, 4, 5 & 6mm

This kit has an excellent mix of tools for working on a car. It contains sockets, spanners (wrenches), pliers, screwdrivers, hex and Torx keys, and a hammer. (Courtesy SP Tools)

### ⅜ inch drive sockets & accessories

* 12-point – 8, 10, 11, 12, 13, 14, 15, 16, 17, 18 & 19mm
* 45 tooth ratchet & flex handle
* Wobble extension bars – 75 & 150mm
* Adaptors – ⅜ female x ¼ male, ⅜ female x ½ male
* Sparkplug sockets – 16 & 21mm
* Torx bits – T27, T30, T40, T45, T50 & T55
* ¼ male adaptor
* Hex bits – 6, 7, 8, 9 & 10mm

### ½ inch drive sockets & accessories

* 12-point – 10, 11, 12, 13, 14, 15, 16, 17, 18, 19, 20, 21, 22, 23, 24, 26, 27, 30 & 32mm
* 45 tooth ratchet & flex handle
* Wobble extension bars – 125 & 250mm
* Adaptor – ½ female & ⅜ male

## SPANNERS

* Combination ring (box end)/open ended – 8, 9, 10, 11, 12, 13, 14, 15, 16, 17, 18 & 19mm
* Ratchet – 8, 9, 10, 11, 12, 13, 14, 15, 16, 17, 18 & 19mm

## GENERAL TOOLS

- Flat screwdrivers – 3.0 x 75, 5.5 x 100, 6.5 x 38, 6.5 x 100, 6.0 x 150 & 8.0 x 150mm
- Phillips screwdrivers – #0 x 75, #1 x 75, #2 x 38, #2 x 100, #2 x 150 & #3 x 150mm
- Allen keys – 1.5, 2, 2.5, 3, 4, 5, 6, 8 & 10mm
- Torx keys – T10, T15, T20, T25, T27, T30, T40, T45 & T50
- Pliers – combination, long nose & curved jaw locking
- Cutters – diagonal
- Adjustable spanner – 250mm
- Hammer – ball pein 450g

To this you would add:
- general purpose electrical multimeter
- fuel pressure measuring gauge
- torque wrench
- a hydraulic trolley jack and jack-stands
- oil filter tool
- spring compressors

To be able to do basic fabrication, add:
- hacksaw
- set of files
- digital calipers
- centre punch
- bench and bench vice

To do bodywork restoration, add:
- bodywork hammer and dolly set

In most workshops, you'd also add welding equipment, a power grinder or linisher, and drill press and drill bits – but we'll get to those in later chapters. If you don't have all of the tools shown above, don't feel disheartened. It took me a long time to build my collection of tools, and it's usually impossible for people to immediately afford all the tools that they'd like.

Rather than laboriously work through how to use every hand tool (how to use a screwdriver, how to use a spanner, etc …) I will now look at some key tools and techniques, including those that are frequently overlooked.

## BENCH VICES

A bench vice is the single most valuable tool you can permanently mount in your workshop. If you can hold whatever you're working on firmly and steadily, you'll find the outcome vastly better – in terms of results, ease of use and safety.

### Types

Bench vices come in a variety of shapes and sizes.

There are two primary aspects that determine a bench

A good quality vice. This one has a width of 115mm (4.5in) and replaceable jaws. (Courtesy Irwin)

vice's size: width and opening depth. The vice is normally specified by width alone (eg a 100mm (4in) engineer's vice). The 'engineer' bit means that the vice is designed for metalworking rather than woodworking, while the '100mm (4in)' refers to the width of its jaws. Vices are available that are fabricated through welding, or cast in only two main pieces. My preference is for cast designs.

Many people like the idea of starting off with a small vice that screw-clamps to the bench. In short: don't. In operation, you'll find it unhappily sloppy – typically, moving around when you try to file or cut an object held in its jaws. These vices are generally only low in strength, so you will also not be able to apply sufficient clamping pressure to the workpiece.

Instead, a 100mm (4in) vice is an excellent beginner's vice. Its cost will be low, but it is still large enough to perform many useful functions. Even a relatively small vice like this should have a base that allows it to be bolted to a bench and

I have a Dawn offset vice like this mounted at one end of my workbench. Mine is slightly larger than the one pictured – it has a jaw width of 125mm (5in) and opens 145mm (5.7in). I use it in conjunction with a much smaller 100mm (4in) vice that's also mounted on the bench. (Courtesy Dawn Australia)

should have removable jaws. Note that this design of vice does not allow long, wide items to be clamped vertically.

Some vices have two additional features – a swivel base and an anvil surface. Both sound really useful, but are not particularly helpful in real-world use. Unless tightened hugely, a swivel base vice will tend to rotate when high forces are applied (eg bending a steel strip to form a bracket), and the 'anvil' part of the vice is hard to access – much better to get an old lump of railway line, or a steel beam, and use that when hammering items flat.

The next size up is a 150mm (6in) vice, and after that 200mm (8in) – the latter getting very large indeed! The best advice is to start with the 100mm (4in) vice, and then if you find you often need a larger vice, to buy a second vice with wider jaws. Keep the first vice mounted – you'll find it useful to have two vices, especially when you have a job carefully set up in one vice and then have a short-term need for another. This occurs more often than you'd think!

When upgrading to a large vice, you might also want to look for one with offset jaws. This design allows you to clamp long items vertically in the vice. Note, though, that offset vices develop lateral 'play' in the jaws more rapidly than conventional vices, so keep this in mind if buying a secondhand vice.

## Mounting a vice
Strange as it sounds, mounting the vice is perhaps the most important aspect of vice-use to get right. If the vice is mounted incorrectly, it will be hard to use and may even be dangerous.

### 1. Height
A vice must be mounted so that when you are standing next to it, the top surface of its jaws is level with your bent elbow. If the bench top is too low, the vice can be raised with a hardwood block. If the vice is too high, you can stand on a

rubber mat or timber board. Having the vice at the correct height will allow much more accurate filing and hacksaw cutting – your fore-arm will be able to move back and forth horizontally.

### 2. Rigidity
A vice must be very securely mounted. This has implications both for the way in which it is attached to the bench, and how secure the bench itself is. Some books recommend the use of coach screws (lag bolts) to attach a vice to a bench – don't! Instead, use heavy-duty bolts, nuts and washers to securely bolt the vice to the bench. Large vices use four bolts, while smaller vices may need only two.

But it's no use attaching the vice securely to a bench if the bench can then 'walk!' The heavier the bench, the better. If you're working with a bench that is not bolted down and is lighter than desirable, make sure that it has a lower shelf on which weighty objects can be placed. A bench that is not bolted to the floor needs to have a mass of at least 150kg (330lb) if it isn't to be easily moved around by objects being manipulated in a 150mm (6in) vice.

### 3. Location
As I've indicated above, vices are normally mounted on heavy benches. But where on the bench? This is a deceptively tricky question. The parting line of the vice is best positioned so that it very slightly overhangs the edge of the bench. In that way, long items can be vertically mounted in the jaws of the vice – and that's even more important when mounting a vice with offset jaws.

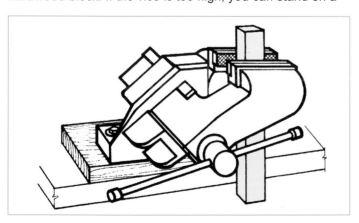

**An offset vice allows long items to be mounted in the vice. Note how in this drawing the vice has been raised by a hardwood block. Vices must be at the correct working height for best results.**

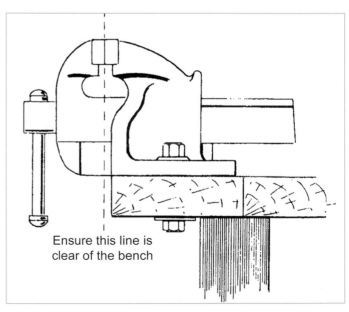

Ensure this line is clear of the bench

**When mounting a vice, ensure that the parting line clears the bench.**

## Using a vice

Vices use hardened and serrated steel jaws to grip the workpiece. These jaws are removable and can be replaced if worn. In normal use, make sure the jaws are kept tight. The serrated jaws keep a tight grip on the workpiece but they also inevitably mark it. So in most situations, the jaws are covered with soft metal. The easiest way of making these covers is to cut some short sections of aluminium angle a little longer than the width of the jaws. These then fit over the vice jaws.

When using the vice, the workpiece is positioned within the jaws and the vice just tightened lightly. The work is then rechecked for location, keeping in mind that you don't want your tool (eg a file) to touch the hardened jaws of the vice. The vice is then firmly tightened by hand. Normally the full width of the jaws is employed to hold the workpiece (or the workpiece is centred) but sometimes the item must be positioned at one end of the jaws. Note, however, that doing this frequently over a longer period will distort the jaws – as a result, the clamping force will no longer be even across the jaws' full width.

### Vice use and misuse

Normally in a section on use and abuse of a vice you'll find points like: never put an extension lever on the vice handle to tighten it, never apply heat to an object clamped in the vice, never use a big hammer to bend bar held in the vice. And so on.

But realistically, over a long period of using a vice, it's very likely that you'll end up doing all of these things – not on a frequent basis, but just when it's required. Along with inadvertent touching of angle grinder discs against the jaws, and normal wear and tear, over time the vice will lose its effectiveness. The thread will get sloppy, the jaws will no longer be parallel (or possibly even at the same height), and, as the vice is tightened, the jaws will change in angle.

It's simply not worth working with a vice that is worn – too many problems then occur. When a vice gets like this, buy another.

## CLAMPS

Clamps are useful in any home workshop. They are used to hold items that are being cut or worked on (eg holding a sheet of steel to a bench when it is being cut) and acting as small, portable presses (eg when pushing back the pistons in a brake caliper).

Clamps shaped like the letters G, C and F comprise heavy cast or rolled-steel frames and square-cut threads. A G- or C-clamp design requires that you manually screw the threaded section to the correct position to apply clamping force, whereas the F-type allows the collar to be slid along the main shaft to speed set-up. Pipe clamps use a length of water pipe or similar (often provided by the purchaser) to

F-type clamps stored but ready for immediate use. These clamps are quicker to adjust than traditional G-clamps but not quite as strong. I bought these secondhand, then cleaned and painted them.

give a very long clamp. These are useful when constructing items like subwoofer enclosures.

Quick release clamps are similar to F clamps, except they use an internal ratcheting and locking mechanism, with the head usually made of plastic. The major benefit of these is that they can be used one-handed, if necessary.

You should have available to you a range of clamp sizes, from those with an opening of only 50mm (2in) to those that can hold items 200mm (8in) thick. Most of my clamps were bought secondhand and then restored – this saved a considerable amount over buying new. On metal clamps, always keep the threads well-greased.

A quick release clamp that can be operated with one hand. (Courtesy Toolstop)

## WORKING WITH NUTS AND BOLTS

The two proper ways of loosening and tightening nuts and bolts are with spanners and sockets. Aside from the different sizes, spanners come in two basic types: ring (box end) and open-ended. Both ring spanners and sockets can be either

This 46-piece socket set uses ½in drive and 6-point sockets. A metric set, it has standard sockets that range from 8-22mm, and deep sockets that cover the 8-24mm sizes. (Courtesy Toolstop)

6- or 12-point. The number of 'points' refers to the number of flats on the inside of the ring spanner or socket.

A 6-point design has a hexagonal shape, and a 12-point design has twice as many internal flats. Six-point sockets and spanners are usually cheaper and are less likely to slip (ie round-off the nut or bolt), but they have a major disadvantage in that they cannot be applied at as many rotational angles. In other words, a 6-point ring spanner may not be able to be fitted on the nut or bolt, as the length of the spanner may foul something. In the same location, a 12-point spanner will fit.

Absolutely vital in successfully working with nuts and bolts is to know the correct hierarchy of use. In other words, which of the available tools should be the first preference?

### Order of correct tool use

### 1st choice – socket
The first choice in doing-up (or undoing) a nut or a bolt should be a socket. If the nut or bolt needs to be torqued to a high value, a 6-point socket should be used. (In normal circumstances, either a 6-point or 12-point socket is fine.) The socket should be equipped with as short an extension bar between the socket and handle as possible – preferably with none. (This to avoid applying a force that tries to lever the socket off the nut or bolt as it is being turned.) The socket can be turned by a ratchet handle, a sliding T-handle or even, for low torque values, a screwdriver-type handle.

### 2nd choice – ring spanner (box end wrench)
The second choice of tool for doing up or undoing a nut or a bolt should be a ring spanner (box end wrench). Where space above the nut or bolt will not allow a socket and ratchet handle to be used, a ring spanner is appropriate. Again, if the nut or bolt needs to be torqued to a high value, a 6-point ring spanner should be used.

### 3rd choice – open-ended spanner
Open-ended spanners are simply great tools ... for rounding off nuts and bolts. Very few open-ended spanners have sufficient strength to undo nuts and bolts that have been adequately torqued. Similarly, very few open-ended spanners have sufficient strength to adequately torque nuts and bolts. Open-ended spanners should therefore only be used when there is inadequate space around the nut or bolt to permit either a socket or ring spanner to fit – and that's a pretty unusual situation.

However, open-ended spanners are very useful in undoing nuts and bolts that have been 'cracked' (ie the initial tightening torque undone) but which are sticky on the threads. For example, a Nyloc nut should be cracked with a socket or ring spanner and then, if it's easier, can be undone completely with an open-ended (or ratchet) spanner.

### 4th choice – adjustable wrench
An adjustable wrench or spanner is even more likely than an open-ended spanner to wreck the head of the nut or bolt. After all, it's just an open-ended spanner with a built-in adjustment mechanism that allows the jaws to spring even further apart when it is being used ... Adjustable wrenches should never be normally used on nuts and bolts. Only in an emergency, where no suitably sized spanner or socket exists, should an adjustable wrench be used.

### (Not the) 5th choice – pliers
Pliers should never be used to do up or undo nuts and bolts. It's as simple as that. The only reason I ever use pliers on a nut or bolt is if it has already been rounded and it's impossible to use a spanner or socket. If you ever

Adjustable spanners (crescent wrenches) are useful in a home workshop but they should not be used to tighten or loosen nuts and bolts at anything but low torques. (Courtesy Toolstop)

Ratchet spanners are extremely useful in a home car workshop. Note, though, that they're generally not as strong as fixed spanners, so often it's best to use the fixed spanner to 'crack' the nut or bolt before then using the ratchet design.
(Courtesy Toolstop)

see anyone using pliers on a nut or a bolt, they simply aren't much of a workman (workperson – gender not implied).

## Tips and tricks

### 'Cracking' nuts or bolts

Most nuts and bolts that have been torqued up are pretty tight. Applying gradual force by a ring spanner or socket will often leave you frustrated and sore – the nut or bolt simply doesn't want to 'give.' The trick is to 'crack' the torque with a sudden, strong force.

On smaller nuts and bolts, hitting the end of the spanner with your cupped hand will often crack the torque. On larger nuts and bolts, using a rubber mallet will perform the same trick. On really big bolts, like those used on suspension components, pushing hard with your foot will often crack them. Now, stamping on a spanner doesn't sound very good workshop practice, but if you're got the car securely on stands and you're already under the car, you can apply a lot of force with your foot in a direction that won't cause the spanner to be levered off.

Another approach, which is perhaps frowned upon in the

textbooks, but which is practically useful, is to use a second spanner to add leverage to the first. For example, imagine you are using a combination ring (box)/open-ended spanner on a very tight bolt, with the ring end of the spanner over the bolt. You can then loop the ring part of a *second* spanner through the distant open jaws of the first, extending the effective spanner length, and so increasing the leverage you can bring to bear.

Finally, don't forget that you can also use hollow tube as an extension to increase a ratchet handle or the spanner's leverage.

### Angled extension bars

Extension bars for socket sets are available with bevelled edges, allowing the extension bar to be out of line with the socket. These are sometimes called wobble extension bars, and, in some situations, they can be an absolute lifesaver. These bars can be bought separately: they're something you should have in your tool box.

### High torque bolts/nuts

If the bolt or nut that you've taken off was very highly torqued, think about why the manufacturer (or whoever previously did it up) made it so tight. If there is an important reason that it should be tight, consider applying Loctite or equivalent locking compound. In many cases, using a locking compound means you won't have to go ballistic on the torque (which saves potential thread stripping or nut rounding), and the fastener will be more secure than before. If you don't have available a suggested workshop manual torque value, and you want to torque to a value appropriate for the fastener's diameter, bolt grade and thread type, there are lots of online resources that give appropriate torque values to use.

### Accurate torquing

Anything involving bearings (plain, roller, needle), important gaskets (eg head gasket) or castellated nuts (ie a split pin goes through the end) should have the bolts torqued as described in a workshop manual. That might mean applying a certain peak torque (as measured by a torque wrench) or by a certain angle of rotation. Note that accurate torqueing requires that the threads and underside of the bolt are first lubricated.

### Direction of rotation

Pretty well anyone who has even picked up a spanner knows that you turn it clockwise to tighten the fastener, and anti-clockwise to undo it. But two points: firstly, these directions are reversed if the fastener is upside-down (and only the other day I saw a mechanic with 30 years' experience get this wrong!); secondly, this convention applies only to right-hand threads. Some special threads are left-hand (and so 'backwards') – for example, left-hand threads are used on some gas cylinders, some fittings associated with rotating shafts and some tie-rod ends.

### Multiple fasteners

If the object you're working on has been held in place with multiple fasteners, never tighten one fastener to full torque before doing up the rest. Instead, torque them up evenly, working your way back and forth across the object. This applies to wheel nuts, cylinder head bolts, tappet cover bolts – anywhere there are multiple fasteners used to hold something in place. And the reason? You want the object to 'bed down' evenly, not end up cocked on one side.

### 'Nipping up'

In most cases, fasteners don't need to be mega-tight. It's one of the most common mistakes a beginner makes – assuming that every fastener has to be as tight as they can make it. In fact, most fasteners can be 'nipped up.' This means tightening the fastener to the point where it starts to resist tightening, and then applying a relatively small but sudden shock to the end of the spanner or socket handle. Sump plugs, for example, should be nipped up – tight enough that they'll never fall out but not nearly tight enough that the thread will be stripped. On the other hand, suspension and brake caliper nuts and bolts should be quite a lot tighter than being just nipped up.

## MEASURING TOOLS

Making accurate measurements is vital in any fabrication work, and also in many car maintenance jobs.

### Tape measures

For non-precise, relatively long measurements, use a tape measure. Retractable metal designs give fine accuracy (don't use cloth tape measures!) and are available very cheaply. The tape should have a moveable end tip – when the tape is pushed up against a surface, the tape measures the distance from the surface but when the tape is hooked over an edge, it measures from the inside of the edge. In this way, the moveable end tip cancels out its own thickness. About the only other point to make is that you should always buy a tape that has a brightly-coloured body – it makes it so much easier to find …

### Steel rules

The next level up of accuracy in general purpose measuring instruments is provided by a steel rule. Again, it's a very

cheap instrument. A typical steel rule is 300mm (12in) long with markings each single millimetre. However, better quality rules have markings over the first 10 or 20mm that are in half-millimetres. (The same applies to the Imperial equivalent – finer markings at the beginning.)

There are two important aspects to keep in mind when using a steel rule:

Firstly, the rule should be placed so that the graduations are physically as close to the work as possible. For example, where possible, the rule should be placed at 90 degrees to the work, so that the graduations actually touch the work surface.

Secondly, a habit to avoid is measuring a work piece from the very beginning of the rule – the end markings of a steel rule may have been damaged (this occurs much more easily than to markings along the inner length of the rule) and it's also much easier to accurately line up two markings, rather than one marking and the end of the rule. However, when you use the rule in this way, remember to subtract the initial numbered amount from the total reading.

A steel rule can also be used to assess the flatness of a surface. Place the rule on edge across the surface then hold the work and the rule up to the light. If the surface is flat, no light will be able to penetrate between the edge of the rule and the work piece.

In any typical home workshop, steel rules wear out – normally because of inadvertent abuse like being trodden on, cut with power tools and so on. Expect to replace a steel rule every few years.

One metre (1 yard) steel rules are also extremely useful – as a marking edge, to check for flatness and of course, to measure lengths. I have a 2-metre long steel rule which has been surprisingly useful, especially in looking at deflections. However, long steel rules tend to be much more expensive than short ones.

When buying steel rules, pick those that are reasonably stiff (some cheap rules are very thin and flexible) and have a dull surface finish. The dull surface finish reflects less light and makes it easier to see the gradations.

### Digital calipers

Digital calipers are nearly as useful as the humble tape measure and steel rule. They are also now so cheap that no home workshop should be without one. Digital calipers are most commonly available in 150 and 200mm (6 and 8in) measuring sizes. Both sizes are useful – the smaller calipers will fit into tighter spaces and the larger ones will obviously measure larger workpieces.

**A humble 12-inch or 300mm steel rule should be in every home workshop. In fact, it makes sense to have a few distributed around the workshop, hanging on nails and accessible whenever needed.**

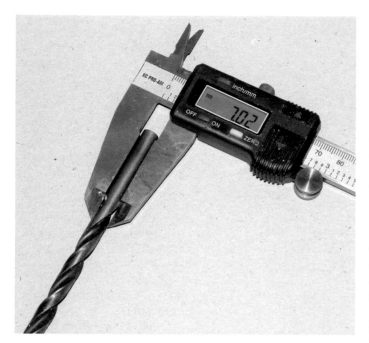

Digital calipers are a must-have. In addition to measuring outside dimensions (as shown here), they can measure inside dimensions and hole depth. I also often use mine to mark centrelines. It's hard to do quick and accurate fabrication, especially of small items, without these.

Calipers are fitted with surfaces to make external, internal and depth measurements. The external ones fit across the work (eg to measure the diameter of a rod), the internal ones measure the width of openings (eg of a cut-out being made in a panel) and the depth provision can be used to measure the length of a drilled hole.

The primary use of a digital caliper is the measurement of the thickness of an item. This measurement requires the use of the external calipers. The instrument is fully closed, zeroed, then the external calipers placed so that they are a snug fit on the item. The reading is then made before the calipers are removed. Used in this way, digital calipers are especially useful in measuring screws and bolts to make accurate sizing of the drill-bit easy, measuring coil spring wire thickness (something that it's important to do very accurately when comparing spring rates) and measuring sheet metal gauge.

Internal measurements with digital calipers should be made only when the opening has parallel sides. If you try to use the internal measuring calipers to measure the diameter of a hole, the thickness of the internal measuring jaws will give an incorrect reading. Instead, in this situation you need to use thinner manual calipers, as described below.

One of my most frequent uses of digital calipers is to mark centrelines. For example, say that you are making a 25mm (1in) wide bracket from aluminium flat bar, and you need to

drill a mounting hole, centred down one end of the bar. If you set the caliper to 12.5mm (½in) you can slide one arm down the edge of the bar, and use the sharp pointy end of the other arm to scribe a light line down the middle of the bar. Do the same from the other edge of the bar and the centreline of the bar will be halfway between the lines. (If you're careful, the lines will be on top of one another and so there will be only one line.) Now decide how far from the end of the bar you want your hole, and set the caliper to that measurement. Use the caliper to lightly scribe this line at right-angles to the ones already drawn. You can now centre-punch the point at which the lines cross, then drill the hole. You will then have a hole centred in the bar and a precise distance from the end. This marking-out approach takes only a few seconds.

With the low price of stainless steel digital calipers, it's not worth buying plastic imitations. It's also not worth hanging on to old-style vernier calipers – the digital readout is much quicker and easier to read, with less likelihood of mistakes being made.

### Manual calipers
There are many situations where a digital caliper cannot be conveniently used – perhaps because the jaws are not long enough to reach the workpiece. In this situation, manual calipers perform well.

The calipers are opened (either by being pulled open against a friction screw or having an adjustment knob unwound) until the jaws of the caliper are a gentle push-fit over the work. The calipers are then transferred to either the digital calipers or steel rule and a reading made of the distance between the manual caliper's jaws. With care, this process can be highly accurate.

Some manual calipers can have their offset jaws completely rotated to allow them to measure both internally and externally, while others are limited to just the one function or the other. With care, calipers last for many decades, and so these are good tools to pick up secondhand. (Note that many older calipers are made from mild steel, so should be oiled to prevent rust.)

### Feeler gauges
Feeler gauges comprise leaves of very thin metal arranged in a pack. They are used for assessing the width of small gaps. If you have an older car with points ignition, they can be used to measure the points gap, and in all cars, they can be used to measure the sparkplug gap. However, their use is not limited to just these pursuits – because they will fit in spaces where no other measuring tool will work, they're very handy to have in your workshop.

## FILES
If you are going to make anything for your car, you'll need to have a selection of files, and know how to use them. Files

are categorised according to two criteria: shape and teeth coarseness.

## File shape

The most common file is a flat file. A flat file has cutting faces that are parallel to one another. In plan form, the file is normally slightly tapered, and has cutting teeth on both minor edges. The main teeth are normally arranged in a double row. This type of file is a general-purpose design suitable for reducing to size, and shape or fit finishing.

A mill file is similar to a flat file, but is thinner and smaller. Its size allows it to be used where a flat file cannot – for example, filling a slot. Mill files generally have only one row of teeth (see below for more on teeth rows).

A hand file is like a flat file but it has teeth on only one minor edge. This characteristic is very important as it allows the filing of a shoulder or internal edge of a square or rectangular cut-out without inadvertently filing away the other surface. If you absorb only one thing from this page, it's this point: that some flat files don't have teeth on their edges!

A round file (called a 'rat tail' if the diameter is very small)

is a very useful file. It is used for enlarging or elongating holes or for filing the inner diameter of tightly curved surfaces.

Another very useful file is the half-round. This is useful for enlarging big diameter holes or for filing gentle curves. Because one face is flat and the other rounded, it's a file that can be used a great deal – one side for normal filing and then without having to put it down and pick up another file, the other side immediately available for filing curves.

A square file has equal widths on each cutting face – it is square in cross-section, but tapers down in size towards the end. This type of file is useful for enlarging square and rectangular cut-outs. In my home workshop, I rarely use this file.

A triangular file is also used rarely. Because it has teeth on all surfaces (including the corners) it is a very easy file to make a mistake with – to inadvertently elongate a corner when using it to file a square opening, for example. However, when a triangular cut needs to be made in a surface (for example, to get a round file started, so creating a half-round opening) it works well. A triangular file can also be used for restoring threads and gear teeth.

## Teeth

Files can be categorised into three tooth types:
* Single cut – the file has a single row of teeth, making for smooth cuts.
* Double cut – two rows of teeth arranged at an angle to each other; most files are like this.
* Rasp cut – raised teeth used for cutting soft materials, eg wood.

Furthermore, the coarseness of files can be classified (in order from most coarse to least coarse) as: rough, coarse, bastard, second-cut, smooth, dead-smooth, super smooth.

## Selecting the correct file for the job

It's very unlikely that you are going to have 20 different files hanging on your wall, ready to go. So most of you are not going to say to yourselves: "Hmm, will I use a second-cut half-round on this or a single cut mill file?" Instead, you're going to grab the file that looks most suitable – even if it really isn't!

When selecting a file, the most important things to get right are the teeth coarseness and the file shape.

Flat

Mill

Hand

Square

Round

Half-round

Triangular

**Files come in a variety of cross-sectional shapes. Each is useful in a home workshop.**

**Differing file coarseness. From left to right: rasp, rough, coarse, bastard, second, smooth, dead smooth, super smooth.**

A file that is too coarse will:

* make very deep scratches which will be hard to remove
* potentially jam on the surface, breaking teeth
* potentially take off too much material – a disaster, because usually you can't put it back!

For these reasons, it is better to select a file with teeth that are too fine, rather than too coarse. The worst thing that can then happen is that it will take you longer to do the job.

The softer the material, the coarser the file you can use – but the quicker material will be removed. A coarse file used on soft plastic can remove a few millimetres each stroke – it takes very little filing before you find that you've gone too far. Therefore, be conservative in the coarseness selection, leaning towards files too fine rather than too coarse.

While tooth coarseness is usually judged just on the appearance of the file (ie teeth per inch), in the case of worn files, the smoothness of the cut may be quite different to the file's appearance. For example, a coarse file with all the edges worn off may in fact act as a smooth file. Therefore, it pays to know your own files, and mentally pigeon-hole them according to their actual performance.

The file shape is critical. When filing a large flat surface, pick the widest file you have available. When enlarging a hole, pick the largest round file that will fit through the hole. In both cases, a smaller file will be much harder to accurately control. However, if you are using a half-round file to enlarge a hole, make sure that the file isn't too large – the edges will dig in and the hole will become ragged.

In all cases, before you start to remove material, run the file gently over the work so you can see where material will be removed. This 'try before you file' approach will allow you to quickly see if the teeth on the edge of a file will unexpectedly take away material, or if the file is too large or too small.

### Filing techniques

Filing's filing, eh? What's there to know? In fact, there are three different filing techniques.

### 1. Heavy filing

Heavy filing is used to remove a lot of material. One hand holds the file handle and the other firmly grasps the other end. The file is moved back and forth across the work, with more pressure being used on the 'push' stroke than the return. A relatively coarse file is used.

When it's possible, mark a line that you're filing towards – and when heavy filing, always stop well before the line. (You'll creep up to the line using light filing.)

When heavy filing a flat surface, it's very easy to round the edges as the file 'rocks' over the work. Therefore, consciously try to keep your arms moving back and forth in a perfectly flat motion, and frequently check the flatness

A good beginner's set of double-cut files. Keep the files, even after they've become blunt, because they'll then be good on soft materials like plastic. (Courtesy Toolstop)

of the resulting surface with a steel rule or square. Where the work can be mounted in a vice, it should be near the height of your elbow as you're standing erect – stooping over the work will make it harder to keep the file moving in the correct plane.

### 2. Light filing

Light filing is used for shallow cuts and when your filing is approaching the marked line. One hand holds the file handle as before, but the other hand holds the end of the file with just the finger tips. The pressure used in light filing is much less than for heavy filing. A finer file is used, and this – along with the lighter pressure – results in less material being removed each stroke. Light filing can be used to 'true' the surface and/or take off just a tiny amount of material.

### 3. Draw filing

Draw filing is when the file is moved over the item with a sideways action. It is a good technique for removing scratches left by the other filing techniques, and can be used to polish the surface. It is also good for fixing square edges that have been rounded by inadvertent rocking of the file.

### File cards

Files that are used on soft materials (eg aluminium or copper) tend to clog. The tiny metal chips caught in the file teeth reduce the cutting efficiency of the file, and can also cause severe scratching of the surface. A file card (a specific bristle brush) can be used to clean the teeth. I simply use a normal wire brush.

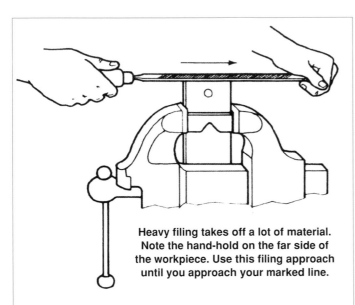

Heavy filing takes off a lot of material. Note the hand-hold on the far side of the workpiece. Use this filing approach until you approach your marked line.

Light filing removes less material and is more precise. Not the changed hand-hold on the far side of the workpiece.

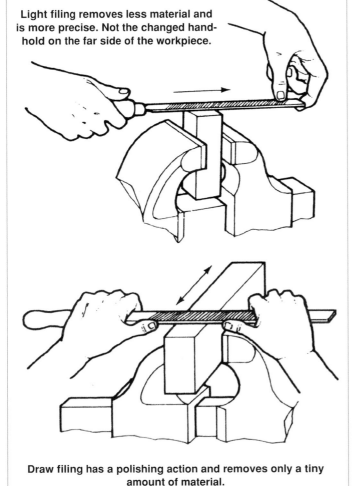

Draw filing has a polishing action and removes only a tiny amount of material.

### File handles

All files should be fitted with handles. Without a handle, a file is difficult to accurately use and the sharp tang is dangerous. I prefer traditional wooden handles that give excellent control. Match the handle size to the file – large handles for large files and small handles for small files.

## TAPS AND DIES

Anyone who makes things will come across threads all the time – most frequently, external threads on bolts and internal threads on nuts.

But, although perhaps you've never thought of it, there are also lots of other threads that you're constantly in contact with. Take an engine inlet manifold – almost certainly there are some bosses cast into it that have been threaded to take bolts. Or even the engine head itself – the manifold bolts to the head using studs screwed into holes tapped in the head.

It doesn't take much thought to realise that having the ability to form threads yourself could be very useful. Sound hard? It isn't – and you need only a cheapish set of hand tools.

### Types

A tool that forms an internal thread is called a tap. A thread-forming tap looks a bit like a threaded bolt, except the thread is not continuous around the diameter (there are

This quality tap and die set also contains a thread gauge. Always keep the taps and dies well-oiled, or they will rust.
(Courtesy Toolstop)

1. BUYING AND USING HAND TOOLS

longitudinal clearance slots) and the thread is formed with a sharp cutting edge. In brief (I'll cover the use of the tool in more detail below) a hole of the correct diameter is drilled in the material, and then the tap used to form the internal thread in the hole.

A die is the equivalent tool that forms external threads – for example, you might have a bar and want a thread on it. In that case, you'd use a die. A die (sometimes called a 'die nut') has cutting teeth on its inside diameter, and forms the thread as it is screwed down over the round material.

A single tap or die can form just the one type of thread – whether that's metric or Imperial, and whether it's a coarse or fine thread. Taps and dies are available in sets, and are usually categorised as metric or Imperial. Obviously, pick whichever type best matches your needs and other tools.

Tap and die sets can cost the earth – or alternatively, be quite reasonably priced. For home workshop use, where the set isn't going to be in frequent use, a medium priced one is appropriate. In addition to the taps and dies, look for the presence of driving handles (usually two sizes for the taps and two sizes for the dies) and, most importantly, a chart that shows the correct drill size for the different taps.

## Uses

About now you might be thinking – that sounds fine, but I'd never use them. And you might be right! What? Using taps and dies requires a certain approach – you need to start thinking: I could tap this to take a thread, and that would that save me a lot of pain. But if you never think that way, you'll never use the tools …

For example, a supercharger bracket that I once made incorporated a belt idler pulley, and it was important that the flanged pulley was adjustable so it could be set at exactly 90 degrees to the surface of the belt. That meant it had to be bolted into place, not welded, so the bolts could be loosened and the pulley adjusted. But there wasn't much room for nuts, and, because of tight access, holding any nuts still while the bolts were being tightened would have been near-impossible. The answer was to tap threads in the steel plate that formed part of the bracket. Neat, easy, quick and strong!

Or, take the job I was working on today – the crush tube through a custom poly suspension bush. The crush tubes are made from 12.7mm (½in) chrome-plated steel bar salvaged from shock absorbers. To hold them in place, a hole has been drilled through the length of the crush tube, and then the hole has been tapped to take two 8mm stainless steel bolts inserted from each end.

The alternative would be to insert a single through-bolt (complete with nut at the other end) but 8mm diameter bolts this long, in stainless steel, are hard to source, and the completed assembly would be heavier (and every gram

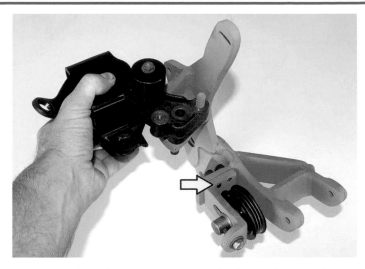

I tapped the threads (arrowed) for these bolts in this supercharger bracket. Working in conjunction with slotted holes, this allowed the pulley to be precisely aligned to the belt. Note how there's not room for nuts on these bolts.

counts in this particular project). Again, having a set of taps made it easy.

But what about dies, then? When do you use them? I don't use them anywhere near as often – it's sometimes easier to buy an off-the-shelf bolt and modify it. However, if you need an external thread in an unusual material or on a prebuilt item, using a die will be the easiest way to get it.

An example of a die that I used in a recent job is when I was fitting air suspension to one of my cars. The suspension height sensor consisted of a sealed, rotary sensor with a ¼in diameter steel arm protruding from it. I wanted to be able to attach a miniature ball-joint to the end of this arm, so I could use a second link to connect to the suspension member of the car. I held the arm of the sensor in a vice and used a 6mm die to form a thread on the end

I threaded the steel arm of this suspension height sensor with a die. This allowed the miniature ball joint to be screwed on. It's prevented from turning by the jam nut.

of the arm, allowing me to screw a miniature female ball-joint directly onto the arm. I made the thread long enough that I could also use a jam nut to lock the ball-joint in place. Neat and effective – and very hard to do without a tap and die set!

### Tapping techniques

There are three key aspects in using taps:

- get the hole size right
- keep the tap square
- use intelligent feel

Let's take them one by one.

### 1. Hole size

As stated earlier, any decent set of taps should include a table that shows the tap thread size and the appropriate hole diameter to drill to take that tap. Getting the hole size right is vital. If the hole is too small, the tap will jam and may break off. (And removing a broken tap is a nightmare!) If the hole is too large, the thread depth will be shallow, and the thread will lack strength.

Rather than rely on the nominal marked size of the drill, always use a pair of digital calipers to measure the actual size of the drill shank. Creep up to the correct hole size in a number of steps, by using a succession of increasing size drills. In this way, the final hole size will be more accurate. Make sure that you are drilling at right-angles to the work – it is preferable to use a drill press.

### 2. Keeping it square

Even if your hole is at 90 degrees to the surface, there's no guarantee that the thread will be – it's easy to cock the tap in the hole and so have an angled thread. That's especially the case if thin plate is being tapped. If the object you're tapping is small and portable, place the tap in the chuck of a drill press and the object beneath it on the table. *Don't switch on the drill press motor!* Instead, get the tap started by turning the chuck by hand. Once the tap is well on its way, you can remove the object from the drill press and place it in a vice, continuing the tapping using the handles provided in the tap set. Another trick is to thread the tap through a nut of the correct thread, and then hold the nut on the surface of the material, centred on the newly-drilled hole. This will start the tap square to the surface.

If you have a lathe, and the workpiece can be mounted in the chuck or bolted to the faceplate, you can use a fixed or rotating centre in the tailstock to line up the tap. This is made easier because the non-threaded end of the tap has a small locating dimple that the fixed or rotating centre will fit into. In this situation, don't use the lathe motor to turn the workpiece, instead lock the lathe and then turn the tap by hand using a suitable wrench.

### 3. Feel

It's easy to wreck the threading of a hole by using a heap of strong-man torque and no brains. Always lubricate the tap (the lubricant to use depends on the material, but you can't really go wrong with plenty of WD40 or cutting oil) and back it out after every few turns, cleaning the chips of metal off the tap and relubricating it. If the tapping effort ever suddenly increases, back out of the tap and clean the chips.

If you're tapping into a blind hole, be aware that without the use of special bottoming taps, the thread won't go right to the base of the hole. Additionally, when tapping into a blind hole, measure the depth of the hole first and mark the tap appropriately so that you know when to stop tapping. If you're tapping through a plate, when the tap is right through, it should spin easily on the thread it has just made.

After the hole has been tapped, clean it thoroughly with compressed air or high-pressure water to remove residual chips. Check with the bolt that the tapped thread works well – if it doesn't, run the tap though a few more times.

### Using dies

Much the same advice applies to dies, but of course the hole size becomes the diameter of the stock. In addition to getting the tap started square (a little easier than with a die because you can see the thread being formed), be careful that the die is centred on the stock. Otherwise, you can get a weird thread where it's deeper on one side than the other – physically weak, and nuts don't like it much either. Most dies are adjustable – make the initial thread with the die open as far as possible, and then squeeze it smaller for subsequent runs, using the adjustment screws built into the drive wrench.

### When not to use

When using taps and dies, keep in mind the strength required in the application. As a very general statement, a hand-formed thread is not as strong as the rolled thread of a bolt or nut. So if a die is being used, the material being tapped to take the bolt or stud should be thicker than an equivalent threaded nut. For example, the steel supercharger bracket described above had the tapped holes made in 10mm (about ⅜in) steel plate. That's pretty strong to hold a few adjusting bolts! The crush tubes made from steel bars used bolts with over 15mm (about ⁹⁄₁₆in) engaging in the threads. But tapping plate that's 2mm (about ¹⁄₁₆in) thick, and then expecting the bolt to hold a big load, isn't wise.

The other aspect to keep in mind is the material on which the thread is being formed. Aluminium takes taps and dies beautifully – but the thread won't be as strong as steel. Plastics can be tapped easily – but again the result won't be very strong. (Might be ideal to hold a trim panel in place, though.)

### Selecting threads

If you are making components with both internal and external threads, and the two parts are going to screw together, make sure you select the matching tap and die! (Don't laugh, I picked up the die one across in the set just yesterday – hmm, big thread, this one …)

If you are making a thread that will be matched with an off the shelf bolt or nut, ensure you have plenty of those bolts or nuts in hand before starting to cut the thread. Why? Well (1) you can check that in fact it is exactly the same thread, and (2) you are certain you can actually buy the appropriate nut or bolt. As an example of the latter, I originally cut an 8mm x 1mm thread in my suspension pivot crush tubes. I had a bolt that matched that thread, and so I wasn't concerned about sourcing more of the bolts – even though the sample bolt wasn't the stainless steel, button head, Allen key design that I eventually wanted to use. After making two of the crush tubes I went off to source some 8mm x 1mm stainless steel, button head, Allen key bolts. Only to find they're impossible to get! 8 x 1.25mm? Oh yes, plenty of those … I had to drill and tap new crush tubes …

### Repairing threads

Both taps and dies can also be used to clean-up damaged threads. In these cases, rather than starting the thread from scratch, the tap or die is used to re-form burred or partially stripped threads. They can also be used to repair threads on bolts or tapped holes where the thread has become corroded. You cannot achieve miracles with this approach; the thread must already be clearly visible on the workpiece, and typically need repair in only one part of its length.

### RIVNUTS

Rivnuts (also called nutserts, blind rivet nuts and similar) can be an absolute lifesaver when working on cars. They allow you to insert special threaded inserts into blind holes, so letting you screw bolts into the holes. To put this another way, they turn empty holes into tapped holes – and unlike using a tap to make a thread, they will work in thin sheet.

Rivnut inserts. These are inserted into a hole and then flared into place using a special tool. The result is a very convenient tapped hole.

### Types of rivnut tools

The most common rivnut tool is very much like a pop rivet gun. By operating the handles, an inner mandrel is drawn upwards with great force. In a pop rivet gun, the 'mandrel' comprises the shank of the rivet, while with the rivnut tool, the mandrel is a threaded buck, onto which the nut insert is screwed. This type of tool is fine if you want to insert only the occasional, small diameter rivnut. However, as with inserting large diameter pop-rivets, if you wish to insert larger diameter rivnuts (eg 8mm (⁵⁄₁₆in) and above) it helps if the rivnut tool has long handles, so that lots of leverage can be applied. However, in smaller sizes, it's not hard to break the bucks by applying too much leverage. (Also remember that in this type of tool, the bucks use left-hand threads where they screw into the tool. You therefore need to rotate the buck clockwise to unscrew it.)

If you are going to be inserting a lot of rivnuts, a compressed air gun is available, which massively reduces the amount of effort needed. However, the best tool for inserting rivnuts for home mechanic use is one that uses a completely different approach. Rather than just pulling on the mandrel through leverage, this type of tool uses a knurled mating surface to stop the body of the rivnut from turning, and then uses the threaded mandrel (working as a screw) to pull the rivnut into place.

I use this screw-type tool for inserting rivnuts. It's much more effective than the lever types (the ones that look like pop rivet tools).

To give you an idea of how much force these screw tools apply, compared to traditional lever-type rivnut tools, I applied a screw-type tool to a rivnut that had already been put into place with a lever tool. Using the screw-type tool, I was able to compress the rivnut substantially more – making it immensely more secure. The only downside of the screw type rivnut tool is that it leaves a mark on the face of the rivnut.

Rivnut tools (all types) use threaded bucks that match the diameters and thread types of the rivnuts being inserted. So if you are going to be inserting both Imperial and metric rivnuts, keep this in mind when selecting the tool.

To use a screw-type rivnut tool, the rivnut is first screwed onto the mandrel. The rivnut is then inserted into the hole and a spanner (wrench) used on the tool to draw the rivnut up tight. Once that is done, the tool can be unscrewed from the captive rivnut.

Rivnuts are available in metric and Imperial sizes. Typical metric kits include a tool and rivnuts in M3, M4, M5 and M6 sizes. Typical Imperial kits include 6/32, 8/32, 10/24, 1/4, 5/16, and 3/8 sizes. Note that larger sizes are available, but these become problematic to insert with just a conventional lever-type hand tool – not enough force can be easily applied, so instead you'll need to use a screw-type or pneumatic tool.

Rivnuts are available in steel, stainless steel and aluminium, and most rivnuts have a serrated collar to allow them to better grip the parent material. Some rivnuts are designed to be used just on very thin sheet. Finally, some rivnuts are designed for near-flush applications, and others work with a slightly raised collar.

### Advantages and disadvantages

The huge advantage of a rivnut is that it can be used to attach a bolt to a panel where access to the back is difficult or impossible. Rather than trying to get a nut on the back of the panel, simply drill a hole of the right size, insert the rivnut and lock it into place. There's your captive nut – it's quick, simple and neat. However, even if you can access the rear of the surface, rivnuts can also be useful where it might just be tricky to hold the nut in place as you screw-in the bolt.

However, there are also some downsides. Rivnuts, especially those inserted with a lever-type hand tool, do not flare a great deal after being inserted – their hold on the parent material is tenuous, especially when compared with a traditional washer and nut. This means that they should not be used in structural applications, or where failure could cause danger. I've seen aftermarket seat brackets bolted to rivnuts inserted through the car's floor panel: this is dangerously stupid. I've also seen heavy electric motor controllers held in place with rivnuts inserted into chassis rails. Again, if you consider the mass of the controller might increase by 40 or 50 times in a crash, this is a silly thing to do. The ability of the rivnut to sustain a load also decreases when the insert is made from aluminium rather than steel.

(Note that some car manufacturers, especially those making aluminium cars, use nut inserts quite widely in their structures. However, if you look at these fasteners closely, you'll see that these machine-inserted rivnuts have a much greater bearing surface on the parent material than a typical hand-applied rivnut.)

To be able to insert a rivnut, the clearance in front of the panel needs to be sufficient that the tool's nose can be inserted. In tight confines, this can be impossible. Where there is insufficient clearance, at a pinch you can insert a rivnut by using a well-greased bolt and washer to pull the rear of the rivnut forward, so expanding it. You'll need to use pliers to stop the rivnut turning initially. The inserted rivnut is also usually a little longer than a conventional nut – again, in some situations, this rear clearance needs to be considered.

Rivnuts are the sort of fastener that you think will be used rarely, or not at all – and then when you have a kit, you find yourself using them all the time! In short, if you're fabricating fiddly stuff, they're great.

### PANEL BEATING

If you are working on an older car, or you want to cut repair costs after a crash, knowing a bit about panel beating can be very useful. The first rule is to buy the right tools.

### Tools

At its cheapest and simplest, buy a panel-beating set comprising a number of hammers (three is good) and some dollies. The hammers will be shaped and weighted to produce only small deflections in the metal with each blow (they'll be light and have heads that typically are slightly curved in end form) and the dollies will have different shapes to match different panel profiles.

Other tools that will probably be necessary include some pieces of hardwood that can be shaped to access difficult spaces, and some smooth lumps of steel that can act as dollies when, again, you're trying to work in difficult spots or with panels that don't match the available dolly shapes. For many panel repairs, you'll also need a comfortable low seat (I use an upturned plastic crate with a cushion) and ear protectors.

If you're dealing with a panel that has been dented so far that the metal has been stretched, an oxy-acetylene welding kit will allow you to shrink these deformed areas, and also allow you to anneal (soften) metal that has work-hardened.

Finally, if you cannot access the rear of the panel, you may need a slide hammer to pull dents out. To temporarily attach the hammer to the panel, you can drill small holes and screw in a self-tapping fitting, or you can use the oxy to braze a screw to the panel and then pull on the head.

### Using a hammer and dolly

The most important panel-beating skill is to develop a proficiency in using the combination of a hammer and dolly. In use, the polished face of the dolly is held firmly against

**This bracket holds in place a turbo heat shield where access to the back of the bracket was impossible. I used stainless steel rivnuts to give durability.**

A beginner's panel beating kit comprising hammers, slappers and dollies. (Courtesy Toolstop)

For a beginner, the most important ingredient in panel beating is patience. To progressively remove even a small dent requires many light blows of the hammer. Don't think of it as needing five or ten medium-hard hits; instead, expect to use 50 or 100 light taps. When you are using the hammer, concentrate on the feel and sound the hammer makes as it hits the metal. If you are using an on-dolly approach, and are moving the dolly around behind the panel, you will be able to tell from the feel and sound of the hammer whether your blows are centred on the dolly (this is important, as often you can't see the dolly).

The sound and feel of the hammer blows will also vary with how hard you are pushing the dolly against the panel. If you push less hard with the dolly, so that it bounces off the metal slightly with each hammer impact, you will progressively push the panel away from the hammer – at the same time as you are also forming the metal to match the dolly profile. If you push really hard with the dolly, you can progressively move the panel in the direction of the hammer.

When working off-dolly, you typically want to miss the dolly by only a small amount. (If you miss the dolly by a long way, it becomes akin to working without a dolly – and you'll much more easily stretch the metal.) The feel and sound of the hammer blows, together with the reaction through the hand holding the dolly, will tell you how close you are working to the dolly.

Be very careful if you don't have the tools to shrink the metal (typically by heating with an oxy-acetylene kit). In that case, getting over-enthusiastic can stretch the metal in the opposite direction to the dent – and without the oxy (or a special shrinking dolly), there's no easy way of remedying the mistake.

A beginner will often not know exactly where they're

one side of the panel, and the hammer is used from the other side.

The hammer and dolly can be used in two quite different ways. In the first approach, the hammer head hits the panel in the middle of the dolly. That is, the metal is sandwiched between the hammer head and dolly. This is called 'on-dolly' work. So when is this approach taken? Imagine a small crease in the panel. The dolly is held against the back of the crease and the hammer is used from the other side, making many light blows. As the metal is pushed against the dolly, it progressively takes on the dolly shape. (So if the dolly is slightly curved, the panel will be shaped to match that curve.)

The other approach is called 'off-dolly.' In this method, the hammer hits the metal slightly to one side of the dolly. In other words, you 'miss' the dolly entirely. However, note that you miss the dolly typically by only a small amount. With off-dolly work, the metal is no longer sandwiched between the dolly and hammer. Instead, the dolly is used to stabilise the panel, and the hammer is used to push the metal back into shape.

Imagine that you have a largish depression in the panel. By working off-dolly you can push the metal back into roughly the correct form – with both the dolly and the hammer doing the pushing. After the large depression is mostly fixed, by working on-dolly you can shape it to exactly the correct profile.

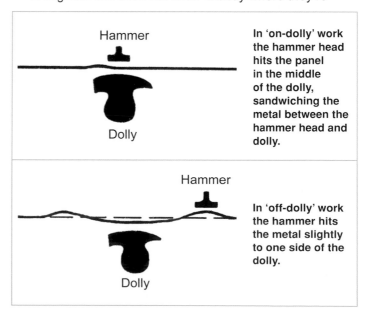

In 'on-dolly' work the hammer head hits the panel in the middle of the dolly, sandwiching the metal between the hammer head and dolly.

In 'off-dolly' work the hammer hits the metal slightly to one side of the dolly.

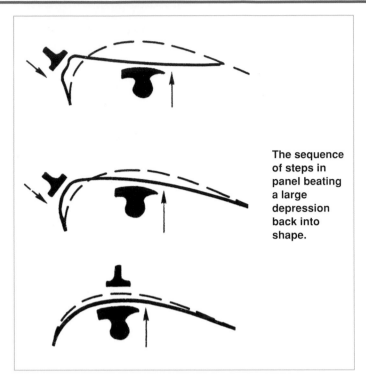

The sequence of steps in panel beating a large depression back into shape.

going. Yes, they're working off-dolly, but is the panel moving in the right direction? At the stage where you're trying to get the panel roughly back into the required shape, run your hand back and forth across it. This will give you a feel for the large high and low areas. Work with many light hammer blows until you can no longer feel large depressions or protrusions.

The next stage in assessing where you are going is to use a paint marker coat and a long sanding board. A long sanding board can easily be made from piece of wood that is smooth and flat and about 300-500mm (12-20in) long. It needs to be about 25mm (1in) thick, and have a width that matches the sanding paper you have available – for example, 100mm (4in). Stretch the sanding paper over the length of the board and use some short screws or brads on the back of the board to hold the sandpaper in place.

The paint marker coat can be from any old spray can – just give the working area a faint spray, let it dry for a few minutes, and then use the long sanding board to go over the area. Use the long sanding board to follow the curves (they're normally convex so you can do this) and you will then immediately see the location of the low areas (no paint removed) and the high areas (paint rubbed off).

You are always aiming to remove as little metal as possible, so it's the knocking down of the high areas which is most important. When you get to the stage that the panel is nearly the right shape, use the pick point of a body hammer to very gently tap down these high points.

You may have started with the idea that you're going to shape every panel to the point of metal finishing (that is, no filler needed at all) but if you're anything like me, you'll find that last step nearly impossible. Instead, aim to have the surface within (say) 0.5mm of perfect. On most panels, using a straight edge will let you assess this. That final step towards a dead-smooth panel will be achieved with body filler. Don't be tempted to use a flap or disc sander to remove those last tiny dents – if the panel is 1mm thick and you've got it within 0.5mm of correct, sanding it smooth will mean you've just halved the thickness of your panel!

### Practice

Lots of people will tell you to practice on an old discarded car panel. But to be honest, I think that this can be more difficult than working on the actual car! That's because in most 'real car' panel beating, the panel is part of the car and so is well supported. Chasing a practice panel around the bench is no fun at all, and the panel also reacts differently because it is unsupported at the edges. Instead, practice on a real car, but start off somewhere where a lower finish is acceptable – say under the body or in a spot that is not obvious.

But take it really, really slowly. The more slowly you shape the metal, the earlier the warning if you're going wrong. Don't forget that on thin body panels, every single blow of the hammer (no matter how light) is moving the metal someplace – you need to be watching and aware of where that movement is occurring.

## BUYING SECONDHAND TOOLS

Buying new tools is invariably expensive – so what about buying secondhand? If you know what to look for, and if in some cases you're prepared to do some restoration work, you can build your collection of tools without having to spend a lot.

### Hand tools

When examining secondhand hand tools, buy on brand and condition. Screwdrivers, sockets, spanners (wrenches), pliers and diagonal cutters can all be very worthwhile purchases. First identify the brand as one that is known for quality, then look carefully and closely at the tools. Are screwdriver tips in good condition? Do ring spanners (box end wrenches) have burred internal flats, are sockets obviously worn, and, with diagonal cutters, can you see light between the closed blades? Do ratchet handles still ratchet – both ways? Especially if the tools were originally sold in sets (eg screwdrivers and sockets) and the set has missing parts, bargains can be had.

Even broken tools can sometimes be worth buying. I recently picked up a pair of multi-grip pliers where the centre pin was missing. The brand was very good, the condition (apart from the missing pin) was fine, and the price was right (free!). Back in the workshop it took just a few minutes with a file and a good quality bolt and nut to make a new pin.

This length of railway line is used as my anvil. Such pieces can be picked up cheaply or even free where railway line maintenance is being performed.

Other hand tools to look out for include hammers, cold chisels, and punches. I also choose to buy used metalworking files, especially those that are round or half-round. But don't a lot of these have worn teeth? They do – and that can often be good when filing plastics or other soft materials. You don't want all the files in your workshop to be worn, but having some that are relatively smooth has proved time and time again to be a real-world advantage.

Drill-bits – especially in the larger sizes – are usually worth grabbing. Large size bits are very expensive new, and even if the secondhand ones are blunt, large bits are much more easily sharpened than small drill-bits.

Finally, Allen (hex) keys are a worthwhile buy, primarily because as sure as night follows day, you'll come across Allen key bolts that are of odd sizes not catered for by a standard Allen key set.

## Bench equipment

Sometimes really great bargains can be picked up in heavy-duty bench equipment – I am thinking here especially of vices. Large vices are expensive to buy new – but are among the most useful of items to have in your home workshop. But before buying a secondhand vice, there are some important checks to make.

Firstly, is the vice made by a reputable company? (However, vice brands aren't like hand-tool brands – there are some obscure brands of vices that are still very good.) Is the thread square-cut? – it needs to be for durability and strength. When they become worn, can the jaws be unscrewed and replaced? When the vice is closed up, does it tighten evenly across the full width of the jaws? (Test this by closing the jaws on a sheet of paper and then seeing if the paper can be pulled out at one end.) When the jaws are tightened, does the moving jaw move upwards or cock itself with respect to the fixed jaw? Don't forget that a thorough wire-brushing and some paint can make even an old vice look like new.

Specialist vices, like those designed to hold pipes, also occasionally pop up. I recently found a great pipe vice on sale at a country market stall. It uses a chain and a threaded adjustment handle to clamp the pipe down in serrated V-jaws. It can be used even on thin wall thickness tube, and holds the tube in place very rigidly.

Another heavy-duty piece of bench equipment that can sometimes be bought secondhand is an anvil. Medium-size anvils, especially if they are obviously old, are the ones to go for (some modern and cheap anvils are pretty horrible to use). If you cannot find an anvil, or your budget is too low, see if you can buy a short length of railway line. The hardened top surface, combined with the vertical web and wide base, work very well as a surface on which to hammer.

## Clamps

Especially if they are getting rusty, clamps (including G, C and F designs) can be excellent used buys. This is one type of product where you may pay secondhand nearly as much as new – but you'll invariably find that the older designs are more rigid and have higher quality threads. Look for heavy-duty castings with strengthening ribs, square-cut threads, and ensure that the clamp closes correctly. Make sure that the threaded section of clamp through which the screw turns is not worn (check this by seeing if there is undue 'play' in the thread).

If the clamps are rusty, clean them up with an angle grinder spinning a twisted wire brush (wear good eye protection, as this combination will always throw wires), and then paint the cast parts and grease the screw threads. Invariably such clamps will come up like new, and be then good for a lifetime of home workshop use.

## Unusual tools

The best tools to buy secondhand are the ones that you'd not source new – primarily because they'd be too expensive for the amount of use you put them to.

A few years ago, I came across some big steel tyre levers at a garage sale. I looked at them and found they were marked as 'forged,' and were made in what was then West Germany. Realising that they would be excellent in any situation where I needed a very strong and long pry bar, I bought them for a pittance. They've been extremely useful, ever since.

Browsing eBay and filtering the results for 'nearest first,' I found someone a short distance away selling a 'brickie's hammer' – a heavy, short-handled square head hammer. I won the auction very cheaply and picked up the hammer (ie no postage charge). When I collected the hammer, I found that a previous owner had arc welded an 'S' on the side of the head. This hammer gets used a lot!

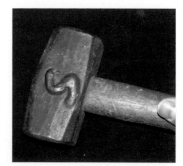

Whenever I use this hammer I am reminded of my cheap purchase by the 'S' the previous owner arc-welded to the head.

**Chapter 2**

# Portable power tools

Portable power tools are cheap and allow you to work fast and accurately. At minimum, you should have at least a few portable power tools, and at maximum – well, the sky's the limit!

## POWER SOURCES

Portable power tools in a home workshop can be powered in three different ways: mains (line) electricity, cordless via rechargeable batteries, or compressed air.

### Mains power

Mains (line) power has been used for portable power tools for many years, and there are significant advantages in using mains power.

First, the tools can be very powerful. As I write this, I am looking at the specifications of a large mains-powered angle grinder. It is rated at 2200W – that's nearly 3hp! The tool weighs 5.5kg (about 12lb) yet has the power of three (small) horses. To run such a tool off 240V mains power will need about 9 amps, and depending where you live in the world, that's likely to be well within the current capacity of a normal domestic system. Even a relatively small electric drill will have around 750W (1hp) of power. With the widespread move to battery-powered tools, don't forget that the sheer power capability of mains-powered tools is unsurpassed.

A mains (line) powered die grinder. Despite the proliferation of battery-powered tools, mains-powered tools still have significant advantages, especially for home workshop use.
(Courtesy Toolstop)

Second, mains-powered tools are cheap. Because they use a simple brushed motor and low-cost control circuitry (even in a variable speed power tool), the cost of the tool is likely to be quite low. Significantly, if you're on a tight budget, the cost of older secondhand power tools can be *very* low indeed. In fact, if the power cable is damaged, often these tools are free! Just replace the cable with one cut off a discarded appliance (I often use junked vacuum cleaner cables – they're long, supple and have a decent current capacity) and you have an electric tool at no cost at all.

Finally, there is a very wide range of power tools available – from grinders to saws to sanders to drills.

And the downsides? Large and powerful tools are

I acquired this mains-powered electric drill for nothing. I gave it a new cord (salvaged from a discarded vacuum cleaner), greased the gearbox and buffed the body with car polish. It will be good now for another 10 or 20 years.

usually heavier than smaller and less powerful tools. If you're doing a job that requires you to use the tool all day, a large and heavy tool can be tiring. You must also always be careful when using a corded power tool. Large sanders can chew through power cables very quickly, and saws can cut the cable as neatly as they can cut through metal or wood! Do not use mains-powered tools outside if the ground is wet. When you are using mains-powered tools in a home workshop, you must use a supply equipped with an earth leakage circuit breaker (a 'safety switch' or ELCB).

If you're just starting out in equipping a home workshop, I recommend traditional mains-powered tools.

### Cordless rechargeable battery power

Rechargeable battery powered tools have undergone a revolution in recent times. Gone are the days when the awful ni-cad batteries were always flat when you wanted to use the tool – and then only lasted a short time anyway. The use of lithium ion batteries and smart chargers means that the batteries retain their charge, have plenty of it, and are fast to recharge.

The major benefit of battery-powered hand tools is that they have no power cord. That means you can just pick them up and start work, without having to find a power supply. You can use them safely outside, and if working on a car, you don't need to worry about the cord getting in the way or getting trapped in rotating components when you start the engine. In some fiddly and precise jobs, like drilling very small diameter holes or inserting small wood screws, not having the pull of a power cord also allows you better control of the tool.

Battery-powered tools are often smaller and lighter than the mains-powered equivalents. However, that's usually because they have less power. There's a reason that

Cordless tools are convenient and safe to use. However, over the long-term their cost is high (batteries need to be replaced) and the viability of the tool is completely dependent on the batteries remaining available. (Courtesy Toolstop)

cordless tools are not rated in watts like mains-powered tools – it's because people would be reluctant to buy them if they saw the figures!

That's not to say that these tools can't do big jobs, but they'll do them more slowly. For example, I have a Makita cordless drill that I love using, but when driving large drill bits, I need to switch the manual two-speed gearbox to its slower speed – and that's very slow. It gets the job done with high torque but low rpm – and since power is torque multiplied by rpm, you can see that the low maximum power the tool can deliver slows down the job. (However, the very slow speed is good when driving large diameter hole-saws.)

Remember also that while battery technology has come a long way, batteries still wear out. If you intend keeping your tools for a long time, that has two implications. The first is that there will be an ongoing cost (forever!) of having a working power tool, and second, you'd better hope that the battery keeps getting produced.

I'd also suggest that you buy tools with brushed (rather than unbrushed) motors. Brushed motor tools are cheaper and you're very unlikely in home use to ever wear out the brushes. Also, the speed control of brushed tools seems better than unbrushed tools – although that may change over time as technology improves.

Unless you have a large budget, I suggest that you buy only a small number of cordless, brushed power tools – for example, a cordless impact driver and power drill.

### Compressed air

Compressed air tools are widely used in professional workshops. The 'zzzzzttt, zzzzt!' of a rattle gun (impact wrench) removing wheel nuts is an example. However, for a home workshop, compressed air tools have a major and significant disadvantage – you need a compressor! And that compressor needs to be fairly large – a small 12V tyre inflater compressor will achieve nothing. A home workshop compressor will require a major expenditure before you can even start to look at buying air tools. (But it gets more complex than that, because the same compressor can be used not just for running compressed air tools, but also for tyre inflation, blowing off dust and spraying paint.)

In addition to requiring a major capital expenditure (a good compressor is likely to cost more than the total cost all of your hand tools), the compressor must also be running if you are to have compressed air available. Compressors are generally noisy, and that can cause problems if you live in an urban area. I live on the edge of a rural village, and even I am reluctant to run my compressor at night. (Quieter compressors are available, but you'll pay more.)

Even a good compressor will tend to lose air pressure overnight, and so cannot be left switched on as it will cycle through the night. The corollary of that is that every time you enter the workshop and flick on power (I always disable power to the complete workshop as I leave it), the

A compressed air tool kit. If you need a compressor for another purpose such as spray painting, compressed air tools make a lot of sense in a home workshop. However, it's usually not worth buying a compressor just to run tools like these. (Courtesy Toolstop)

compressor will start-up. And if you leave the compressor switched off, when you decide to use a compressed air power tool, you'll need to wait for the tank to come up to pressure! If that all sounds negative, it's because what suits a professional workshop where air tools are being used all day does not necessarily suit a home workshop.

And the air tools themselves? These tools are often very light in weight and yet still powerful. Most achieve this by using high rotational speed and low torque. For example, an air angle grinder or die grinder rotates very fast indeed. Other air tools use gearing to provide lower speeds and higher torque – for example, the rattle gun (impact wrench) mentioned earlier. Because of their simple construction, many air tools are cheap.

Air tools always need to be connected to the air supply, and so like mains-powered tools, have an umbilical cord to drag around behind you. However, in contrast to mains-powered tools, the air hose is likely to be stiff and heavy.

Think carefully of what you would like to do in your home workshop before going the 'air' route. If you want to spray-paint a car, do a lot of work on heavy high-torque fasteners like suspension bolts, want to be able to bead-blast, or would like to use the cheap high-speed tools like die-grinders, air may be the answer.

## THE TOP PORTABLE POWER TOOLS
### Drill
A drill is a portable power tool that you'll use all the time. You should aim to get one with a chuck capable of taking at least a 12.5mm (½in) drill bit. The chuck can be the traditional type that uses a chuck key, or it can be a keyless chuck that is tightened by hand. A drill with variable speeds is useful. That can be achieved by either a mechanical gearbox changeover (slow and fast speeds) and/or an electronic variable speed trigger control. In addition to drilling holes, if you have only a small amount of workshop equipment, you can use a drill for grinding, for wire-brushing and for operating a sheet metal nibbler.

Using a power drill is straightforward but there are a few tips to keep in mind.
- Always centre-punch the work where you want the hole to be located. It's impossible to locate holes accurately by 'freehand' drilling – the drill will wander and your hole may end up in the right spot – or more likely, won't.
- If drilling large holes, start with a smaller drill bit and work your way up in steps. That's especially the case when making large holes in thin sheet – using the large drill bit from the beginning will give you a hole that's not round.
- Learn to feel and listen to what the drill bit is doing, and when it's about to break through the workpiece, reduce both pressure and speed.
- Keep drill bits sharp – a blunt drill is slow and will often

A high quality and compact corded drill. A drill is one of the most-used tools in a home workshop. (Courtesy Toolstop)

develop a burr around the drilled hole.
- If you have a drill with a chuck, attach the chuck key to the cord so that you don't lose it.
- Use a drill speed that suits the material. The harder the material, and the larger the drill bit, the slower the rotational speed should be. Err on the side of too slow a speed rather than too high a speed.

### Angle grinder
An angle grinder is another tool that you'll use frequently, especially when fabricating. A 115mm (4.5in) grinder is small enough that it can be manipulated into difficult spots, but large enough to still achieve good outcomes. Lower cost angle grinders will have a lot of vibration and be less

An angle grinder is a very versatile power tool. An angle grinder can be fitted with a cutting disc (as here), grinding wheel, flap sanding wheel or wire brush. (Courtesy Toolstop)

This accessory screw collar allows you to change angle grinder discs quickly and easily, just by hand. The lever flips back down after you've fitted a new disc.

screwdriver is also invaluable when inserting self-drilling and wood screws. Ensure that the tool will lock different bits into place (a quick-release collar is often used) and that the tool does in fact have an automatic impact function it implements at high torque levels. Don't forget to buy a wide selection of bits (including nut drivers) to go with the tool.

This battery impact driver will speed-up a lot of repetitive operations when working on a car. Note that the battery and charger need to be bought separately. (Courtesy Toolstop)

durable, but otherwise will achieve the same outcomes as more expensive grinders.

In addition to grinding metal, an angle grinder can also take thin cutting discs, allowing you to cut sheet and bar, and can power wire brushes, allowing you to clean up metal. Note that the greatest load on a grinder occurs when driving a large wire brush: very cheap angle grinders may then go up in a cloud of smoke! In addition, flap-type sanding wheels are available that allow you to sand flat and curved surfaces.

An angle grinder, especially when driving a wire brush or thin cutting disc, is a dangerous tool. Angle grinders typically spin at around 11,000rpm – very fast. Ensure that the guard is positioned correctly, and you must wear ear and eye protection. When using an angle grinder, use only the area of the blade from 11 o'clock to 4 o'clock, as viewed from the operator's side. If the blade is grabbing and threatening to yank the tool from your hands, you are doing something very wrong. Do not place any part of your body in line with the spinning disc, and never apply any lateral loads to thin cutting discs. When grinding and cutting, ensure that the sparks cannot come into contact with flammable materials – and that includes rags. Finally, when changing discs or blades on mains-powered units, disconnect the power.

## Impact driver

The best impact drivers are small, battery-powered designs. The lack of a cord makes them more nimble, and the impact mechanism means that they will tighten and untighten smallish nuts and bolts to quite high torques. An impact

## Jigsaw

A jigsaw is useful in cutting thin sheet material. If placed against a guide, a jigsaw can cut in straight lines, although its design purpose is to allow you to cut curves. Because the reciprocating speed of the blade needs to vary with different materials, a jigsaw should have an electronic variable speed control. Jigsaw blades are available in metal-cutting and wood-cutting versions – I often use a 'wood' blade when cutting thicker aluminium sheet. When cutting metal with a jigsaw, ensure that you lubricate the blade (auto trans fluid works well), and use a blade of sufficient tooth fineness that it has at least 1.5 teeth in contact with the material. I once cut out a 10mm (about ⅜in) thick aluminium intake manifold flange plate, using just a jigsaw and an electric drill – and a lot of patience.

When using a jigsaw, ensure that both of your hands are always in full view. It is not good practice to hold the workpiece with one hand and operate the jigsaw with the other. Instead, the workpiece should be securely clamped in a vice or to a workbench. If using a mains-powered jigsaw, loop the cable over your shoulder so that there's no possibility of the cable being cut through. Don't forget the blade moves up and down during the cut – there needs to be enough clearance behind the material being cut so that the blade will not foul. This is a trap when, for example,

A variable speed jigsaw can be used to cut in straight lines (using a straight guide) or curves. With appropriate blades, it will cut both timber and metal. (Courtesy Toolstop)

cutting an inner door panel – you don't want the blade striking the inside of the outer door skin …

### Heat gun

A heat gun is another useful tool. Select one that has two settings (hot and hotter). The colder of the two settings will be used when drying paint, when heating rubber hose so that it will better slide over fittings, and similar uses. The higher setting is useful when softening plastics for bending (eg acrylic sheet) or flaring (eg PVC pipe), and for activating heat-shrink on wiring. Heat guns are cheap, so ensure you buy a powerful one like 2000W.

Heat guns are more fragile than they look – dropping one on the ground is likely to damage the heating element, especially if it's dropped after it's been operating. It's easy to get burns from heat guns, either through touching the hot end after it's been operating, or, more insidiously, from the hot air itself. The latter can be quite painful, and not be apparent until a few hours after using the heat gun.

### Sander

Sanders are available in orbital, random and belt configurations. If you intend to do car bodywork, you're best placed buying a random sander, that takes a flat sheet and moves it in a random pattern. An orbital sander, on the other hand, moves the sandpaper in ellipses, while a belt sander moves in just the one direction. In all cases, having a variable speed control is useful. Before selecting a power

A heat gun has lots of uses in a home workshop. Here the (then younger!) author is using a heat gun to soften a rubber elbow being used as part of intercooler plumbing.

An air-powered 150mm (6 inch) orbital sander. It's suitable for bodywork final sanding, finishing, and pre-polishing processes. (Courtesy Toolstop)

sander for general use, carefully consider whether you'd be better placed spending the money on a fixed linisher/belt sander, or using sanding flapper wheels in your angle grinder. Whilst I have portable orbital and belt sanders, I don't use them very often.

## Rattle gun

Rattle guns (sometimes called impact wrenches) are available in battery and compressed air versions. They are particularly useful for undoing very tight nuts and bolts. (Doing them up should not be done with a rattle gun – instead, use a long-lever handle and a socket.) The maximum torque capability goes up with cost; unless you can afford a powerful rattle gun (eg 800Nm of 'bolt busting' torque), it's probably not worth buying one. To use a rattle gun, you also need to buy impact sockets – don't use your normal sockets as they will break.

An air-powered die grinder. If you already have an air compressor, a die grinder is a cheap buy. (Courtesy Toolstop)

## Die grinder

Die grinders take 6.4mm (¼in) shaft burrs and other cutting tools. They rotate at very high speed and are useful for grinding in confined places – for example, inside a tight bracket that you're making. Traditionally, die grinders were powered by compressed air, but mains- and battery-powered die grinders are now also available.

Die grinders can be hard to control until you're well-practised, so keep this in mind if you're new to power tools. Good quality cutting burrs are expensive. You must wear excellent eye protection when using a die grinder – bits of metal fly off at very high speed. Do not use cutting bits in a die grinder that were not designed for high speed use – they can shatter.

A high-torque, air-powered rattle gun. It's ideal for undoing fasteners like suspension nuts and bolts. (Courtesy Toolstop)

## WHAT TO BUY

If you're on a tight budget, I suggest that you buy a mains-powered drill, a battery-powered impact driver, and a mains-powered angle grinder.

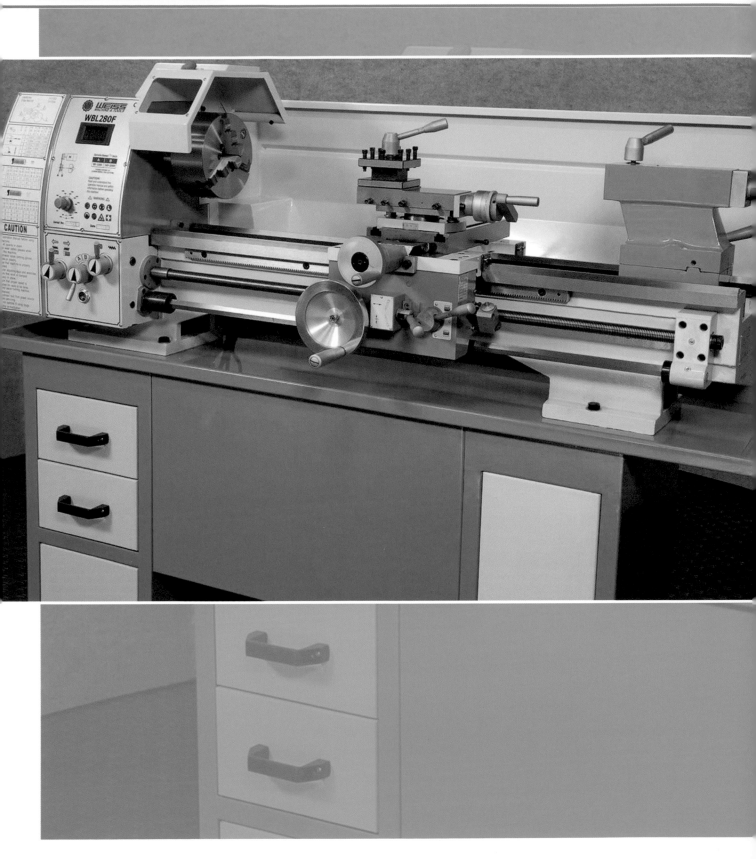

**Chapter 3**

# Major fixed tools

More than any other tools in a home workshop, it's the major, fixed tools that make the difference between a workshop than can achieve great results, and one that's really just a place in which to store a car.

In this chapter, I'll cover all the major tools you might find in a well-equipped home workshop, excepting welding gear, which I'll cover in Chapter 8.

## DRILL PRESS

In the previous chapter I covered portable power drills – so why the need for a drill press? A drill press (sometimes called a pillar drill) has two major advantages over a portable electric drill.

First, it lets you drill holes at fixed angles – usually (but not always) at right angles to the material. If want to drill through a piece of square tube, and want the holes to be positioned on the centrelines of both sides, you'll need a drill press if you're to do it in one operation. If you want to drill a piece of steel bar prior to tapping a thread in it, you'll need a drill press if you want that thread to be square to the surface.

Second, a drill press is often more powerful than a hand drill, and allows you to select from a wide range of speeds. If you need to drill a 25mm (1in) hole through thick steel, a drill press is the tool to do it. (You could also use a mill, which I'll get to in a minute.)

Small and cheap drill presses have around 350W of power and five speeds, selected manually by swapping a belt between pulleys. These presses have adjustable height working tables, with the adjustment by a simple friction

collar and threaded release. Despite the low power, they're still capable of medium-duty drilling jobs – the induction motor and 650rpm gearing giving more grunt than you'd expect. Having a drill press of this type is much better than having no drill press at all.

However, if you can afford it, it's better to select a machine capable of heavier drilling jobs. For example, this might comprise a 16-speed drill press with a 550W motor. The platform on a drill press of this type is likely to be adjustable via a rack and pinion, making it quicker and easier to change the height. There will also be an adjustable depth stop, and may even be a digital speed readout and light. The lowest speed is likely to be something like only 120rpm – but with enormous torque. With a larger tool like this you will physically be able to fit bigger workpieces on the table. Some pillar drills even allow the head to swivel away from the base and table, allowing the end-drilling of long objects.

Larger drill presses are usually equipped with spindles having a standard taper – eg, M2. This allows the chuck to be easily changed should it become damaged, and allows the use of large drill bits designed to be held on a taper rather than in a chuck. Drill presses of this size are available in floor- and bench-mount models.

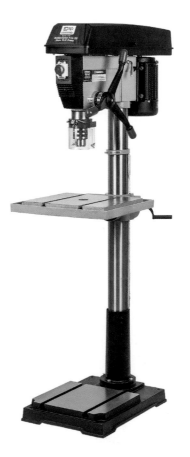

This is a heavy duty, floor-standing drill press. The chuck can take drill bits up to 20mm (¾in) in diameter and a 1100W motor is used. There are twelve speed settings, giving a range of 150-2450rpm. (Courtesy Toolstop)

Small drill presses like these are sold under a variety of brand names. They're suitable for only light work, but they're much better than having no drill press at all. (Courtesy Toolstop)

When you buy a drill press, buy a vice to suit it. Drill presses – even small ones – develop plenty of torque, and so if you're hand-holding the item being drilled, it can easily be yanked from your grasp. This can injure both you and the workpiece. I usually leave the drivebelt a fraction loose so that if the drill bit jams, the belt slips. You don't want the belt slipping in normal drilling operations, so some adjustment might be needed to reach this happy medium. Wear eye protection when working with a drill press. If you can afford it, it's better to buy a more heavy-duty machine, with, for example, 16 speeds and a 550W motor.

## BELT SANDER/LINISHER

While most people would recommend buying a grinder as the next step after a drill press, I'd suggest a powerful bench or floor-mounted belt sander would be much more useful. However, the fly in the ointment is the word 'powerful.'

Many bench-mounted belt sanders are sold for hobby woodworking use, and are severely lacking in power. For example, I once bought a 375W sanding machine that had a 100mm (4in) wide belt and 150mm (6in) sanding disc. It was absolutely useless, and so I sold it (hopefully the buyer wanted to sand just balsa wood!).

One better approach is to buy a 6in or 8in bench grinder, and then add a multitool attachment. These attachments add a 50mm (2in) wide belt and a 175mm (7in) sanding disc to one end of the grinder, replacing one of the two grinding wheels. The attachments are strong and will work well for a long time.

However, there are two disadvantages. The first is that not only do you need to buy the multitool, you also need to buy the bench grinder! The second is that the power of the machine is dictated by the power of the grinder – typically 300-600W. And if you're using a 300W grinder, you're back to the issue that I covered earlier – just nowhere near

A small belt sander like this is too light-duty to be effective in a car home workshop. Motor power is only 375W. (Courtesy Terratek)

enough power. The 600W grinder will be better – and, in fact, if you're on a budget, I'd suggest a 600W grinder and multitool attachment. Still not cheap, but much better than the 375W woodworking machine. I used a multitool on a 6in grinder for years.

However, to gain a machine that can grind and sand metal quickly and well, you need a belt sander designed for metal (these are often called linishers). The one I use is a tool I'd strongly recommend – a Radius Master. It uses a 50 x 1220mm (2 x 48in) belt that wraps around a series of rollers in a roughly triangular shape. One stretch of belt between the rollers is backed by a platen, allowing surface grinding (ie flattening). Another length of belt is not backed by any surface, so can be used to sand slight curves.

The main rollers on which the belt rides comprise 2in (50mm), 3.5in (89mm) and 8in (203mm). These different size rollers allow the grinding of different radii. For example, the large roller is excellent when hollow grinding, while the smaller rollers allow you to get into tight spaces. The smallest roller is located on a rotating mount. Pushing in a pin and rotating this mount allows the selection of another two rollers, the smallest of which is just 1in (25mm). This smallest roller is excellent for notching (fish-mouthing) tube. Note that no adjustment of the belt is necessary when making this roller change. Each of the rollers can be covered as required by swivelling guards.

The head of the Radius Master rotates so that you can bring whichever wheel you want to the front. There are also locking positions to have the platen vertical or horizontal. Furthermore, if you wish, you can rotate the whole machine so that the belt is moving horizontally, not vertically. Powering the belt is an industrial-sized 1.5kW (that's 1500W) direct-drive electric motor.

The Radius Master is a joy to use. Probably the single best aspect is its power – and so the speed with which jobs can be done. With a multitool, you get used to applying only

An 8-inch grinder fitted with a Multi Tool belt and disc sanding attachment. The Multi Tool works very well, but always match it with the most powerful bench grinder you can afford. (Courtesy White International)

very light pressures, or grinding only one part of the job at a time. Push that machine any harder, and it will stop! With the Radius Master, there's so much power that you can just chomp through the material at an incredible rate. It is far faster than an 8in bench grinder in shaping metal, and can of course do things that no grinder can do.

The Radius Master's grunt also means that it can drive power-hungry belts like those made of Scotchbrite material. These belts, that are excellent for polishing (eg cleaning-up aluminium), are stiff, and so require lots of power to drive them. The Radius Master will drive even a coarse Scotchbrite belt at speed – and still have enough power left over to do the work, even at high working pressures.

The Radius Master is one of the most expensive machines in my workshop, but it is also the single most frequently used, permanently-mounted power tool. (Note that high quality, powerful linishers rarely appear secondhand.)

An 8-inch, 600W bench grinder. Consider carefully whether it's worth buying a grinder like this, then later upgrading it with a multitool belt sanding attachment – or instead going straight to a heavy-duty belt linisher. (Courtesy Metabo)

I use a Radius Master as my belt sander and linisher. It is a wonderfully powerful and effective tool that I highly recommend. The downside is that it is expensive. (Courtesy Radius Master)

## BENCH GRINDER

As I described above, a really good belt sander can do almost everything a bench grinder can do – faster and better. (The exception is sharpening drill bits and other tools, where a fine stone is often needed.) Therefore, if you have bought an effective belt sander/linisher, I wouldn't bother buying a bench grinder. However, if you go down the multitool path, where one of the grinding wheels is replaced by the multitool, ensure that you keep the finer grinding wheel and replace the coarse one. And, if you'd like a good belt sander but cannot afford it, a bench grinder will still allow you to shape metal, sharpen drill bits and so on – at low cost. Bench grinders are often sold secondhand, so bargains can also be had.

## FRICTION CUT-OFF SAW

By far the cheapest approach to chopping metal is to buy a friction cut-off saw. The most common and lowest priced of these machines use cut-off discs that are 355mm (14in) in diameter. (Note that more serious versions of these saws use larger blades of 400mm (16in) diameter. They are a lot more expensive though.) Friction cut-off saws are cheap, will abrade their way through nearly anything (even, if you have enough time and spare discs, thick steel sections), and can usually be adjusted to do angles – eg 45-degree cuts.

A friction cut-off saw is a cheap way of cutting metal tube, rod and bar. It's noisy, not very accurate and the material gets very hot, but it's a lot easier than using a hand hacksaw! (Courtesy Bosch)

But they also have some significant disadvantages. The first disadvantage is that they are incredibly noisy. If you work in an urban area, that immediately limits how much you can do at night. You should also always wear good quality ear muffs – and get anyone else nearby to wear them as well. The second disadvantage is that a friction cut-off saw is not a precision machine. They used pressed steel baseplates (rather than castings) and invariably the blade will not be perfectly square to the baseplate – and there's no adjustment. A final disadvantage is that the material being cut gets hot – really hot. If you remember that, it's not a huge problem – but you certainly will need to quench the material after cutting it.

Advantages? They can cut hardened and stainless steels – materials that normally cause problems (or fast blade wear) in other saws. Another advantage is that they can cut tubes largely square at the lowest cost. For example, if you are making a car exhaust or the plumbing for an intercooler, a friction cut off saw is the easiest way of cutting the thin-wall tube square, so that it can be butt-welded together with no gaps. (Yes, you can use a hand hacksaw, but take it from me – for its very low cost, a friction cut-off saw is worth buying for even just one of these jobs.)

No-name friction saws start quite cheaply, while a brand name, good quality machine can double or triple the price. Over the years, I've used both the no-names and the brand-names: the brand-names last longer before the chassis breaks or the motor fails, but the advantages of the more expensive units are not any greater than that.

## BANDSAW

Think 'bandsaw' and most people think of a vertical bandsaw, where the blade travels vertically. But metal-cutting bandsaws are, these days, most often horizontal designs. In this approach the bandsaw blade, its associated wheels, and the motor, are all mounted on a pivoting assembly. The whole assembly pivots downwards (rather like the friction drop saw) as the blade chews its way through the material.

A major differentiator of horizontal bandsaws is the feed mechanism. Some use the weight of the assembly (partially balanced by a spring) to apply downwards pressure on the blade, while others use a more sophisticated adjustable hydraulic system. Some bandsaws also automatically turn off the motor when the material has been fully cut, while others stay running until the operator switches off the machine.

Most bandsaws are highly adjustable – ball-bearing rollers guide the bandsaw blade, and these rollers are adjustable to keep the blade square to the work. The tension of the blade can be adjusted, and the amount of blade that is 'free' (ie not guided) can also be changed to suit the size of the work.

Making angled cuts is done in one of two ways – it

A horizontal metal-cutting bandsaw. I have a very similar model to this one except mine incorporates a swivel base that allows angles to be cut with good precision. (Courtesy Toolstop)

depends on the design of the machine. The simplest machines use a vice that changes in angle, allowing the angled cut to be made. Two downsides of this approach exist. Firstly, the moveable vice jaws are usually not as rigid as fixed jaws, and so inaccuracies can occur with sloppy positioning. Secondly, with this approach, the material must also swivel if it is to be angle cut. If the material is long, clearance in your working space can become problematic. The more complex (and expensive) machines use a swivel head to make angle cuts. The vice can then be more rigid in design, and there's no need to have material going off at odd angles within your workspace – the material is always on the one plane.

Bandsaws down at the home workshop end of the field are typically low in motor power – eg 375W of induction motor power. (To put this another way, they use efficiency rather than brute force.) Even a low power bandsaw works very well, so you don't need to aim for the most powerful. Bandsaws vary substantially in the maximum size of material they can cut. Therefore, if you are looking at cutting, for example, a 100mm (4in) tube at 45 degrees, check the 'size' specs carefully. If set up carefully, a bandsaw can be sufficiently accurate for almost all purposes that don't require machined (or ground) surface accuracy.

Bandsaw blades vary in teeth per inch, construction (cheaper carbon steel versus bi-metal) and quality. It is not hard to destroy a bandsaw blade, so if you're to achieve a good blade life, care needs to be taken in terms of selecting speed, blade type, feed rate and lubrication. Talking of lubrication, many bandsaws have pressure-fed cooling of the bandsaw blade via a pump and nozzle. If the saw does not have this, cooling lubricant should be manually applied via a suitable fluid in an oil can.

With the range of pricing available, the versatility and the ability to set up the saw for accuracy, a bandsaw is a good step up over a friction cut-off saw.

## GUILLOTINE

A hand guillotine (sometimes called a hand shear) works rather like a large pair of scissors. A heavy-duty blade, typically operated by a long handle, moves closely past another (fixed) blade, so shearing the material. A hand guillotine can be used only with solid material (so not pipe, tube or RHS), and typically is able to work with material that is relatively thin (eg less than 5-6mm (¼in) mild steel) and about 150-200mm (6-8in) in width, depending on the model.

However, these specs are a bit misleading. Typically, if you are using the full width of the shearing jaws, you'll be cutting material that's quite thin – eg aluminium or steel sheet. On the other hand, if you're working with flat bar, you might be able to go to full 6mm (¼in) thickness – but only for material perhaps 25mm (1in) wide.

Hand guillotines are quick to use and are typically not that accurate. They're perfect though for cutting off 25mm (1in) bar stock (eg to make a bracket), for shearing small pieces of aluminium sheet to size, and the like. The major benefit is the speed – it takes only moments to make the cut.

A small bench-mounted hand guillotine (sometimes called a hand shear) is a useful tool for cutting small pieces of metal quickly to size. (Courtesy Kaka International)

If you do lots of 'little' work – making brackets, cutting sheet steel patches for bodywork repairs, working with small pieces of aluminium sheet – then a pair of hand shears will be very useful in your home workshop.

Note: while not dangerous in terms of potential ear or eye damage, a pair of hand shears will easily remove fingers. Always lock the hand guillotine when it is not in use – especially if children are ever in the home workshop. Locking the tool can normally be achieved by placing a pin (I use a bolt and hand-tightened nut) through the appropriate holes in the tool.

A foot-operated guillotine also cuts sheet metal to size, but compared with the hand guillotine, the material is thinner and the sheet size larger. In addition to the hand guillotine, I also have a foot-operated guillotine that I typically use when cutting thin steel and aluminium sheet. A foot-operated guillotine is a larger and more expensive tool than a hand guillotine. If you're unsure of how often you'd use a guillotine, buy the hand-operated one first.

## LATHE

A lathe is a tool that fits into a different category to the power tools mentioned so far. Basically, it needs a higher level of skill to produce accurate, well-made work. But I should qualify that. To cut screw threads, to machine standard tapers, to make items with an accuracy that gives good push-fits – these need a high degree of skill. On the other hand, to use a lathe to face off a length of bar so that the end is square, to turn down the diameter of bar stock to make it a suitable size for running a die over it to form a thread, or to centre-drill a disc or shaft – all of these can be learnt quickly and fairly easily. I write this because it's easy to assume that high-level skills are automatically invoked if you spend the money on such a machine!

There are plenty of on-line retailers that sell smallish lathes, usually minus the stand. However, they're often not particularly high-quality units, and by the time you add a four-jaw chuck, face plate, tailstock chuck and stand, the cost goes up substantially. These lathes usually have a distance between centres (ie the longest object that can be continuously machined) of 250mm (10in), and a swing (the radius of the largest cylinder that the lathe can machine) of 140mm (5½in). So they're only small lathes – but that doesn't mean they're useless. In fact, most of the work I do on my lathe could be done on this sized machine.

Note that even smaller lathes are available. These lathes often use extruded aluminium beds and are tiny – just about small enough to fit in a shoe box – and light to match. They use 12V motors and the power supply and motor speed control need to be provided by the purchaser. However, these lathes are not designed for machining steel – they're primarily designed for soft materials like aluminium and plastics. Users of these lathes say that you can machine

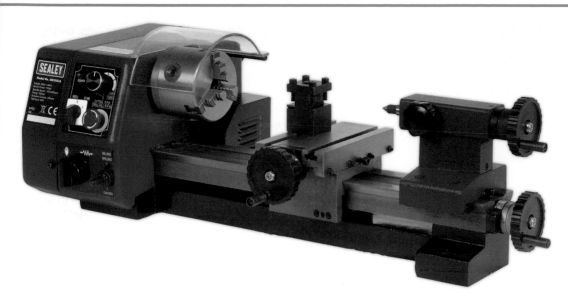

This is a very small lathe. It is better than nothing, but you should probably put the money towards a larger, possibly second-hand lathe. (Courtesy Toolstop)

steel, but you need to take very small cuts indeed. And of course, these lathes can rarely cut screws. In an automotive home workshop, these lathes are best avoided.

Online, and at machinery retailers, you'll also find plenty of new medium-sized lathes. These have a distance between centres of about 500mm (20in) and a swing of 230mm (9in). These lathes represent a good compromise between capability, accuracy and cost. Larger lathes are of course available, but in a home workshop they start taking up a lot of space and are expensive.

My first lathe was a cheap 1940s US-made Southbend.

This is also the path that I suggest new lathe acquirers take – buy a low-priced, used machine. Realistically, a beginner won't have the level of expertise to assess the condition the lathe is in (unless you take along an expert), but the normal tactics of buying secondhand goods (how trustworthy is the seller, do the goods look well maintained, what's their history) will hold you in good stead. Note that I'd be wary of buying an ex-school lathe, simply because you can be confident that every abuse ever handed out to a lathe has been gleefully done so!

For a low price you should be able to get a lathe that will

A good quality lathe suitable for home workshop use. (Courtesy Weiss Machine & Tools)

have a feed for parallel turning (indicated by a powered screw thread that runs the length of the bed) and you'll probably also get 3- and 4-jaw chucks, a dead centre and some tools. Don't worry if all that's another language: any beginners' reference book on lathes will explain all.

In addition to the lathe and its attachments, you will also need to buy suitable lathe tools. Carbide-tipped tools are best for a beginner, although if you are machining primarily aluminium and plastics, and have a bench grinder, you can form your own tools from tool steel.

But what would you actually use a lathe for in car maintenance and modification? Some examples of what I did with that first lathe are indicative.

This adaptor ring (arrowed), that allowed me to change the location of the wastegate actuator on this turbo, was made on my lathe.

### Turbo wastegate mounting

The first major job was the mounting of the wastegate actuator on a turbo. The turbo was one from a Subaru Liberty (Legacy) twin turbo B4. In the Subaru application, the exhaust feed comes up from below, whereas in its new application, the exhaust manifold needed to come in from above. The turbine housing was rotated to get its entrance in the right place, in turn necessitating a new mounting arrangement for the wastegate actuator – it normally bolts to cast-in bosses on the compressor cover.

To provide a new mounting surface, I used the lathe to turn a thick aluminium ring. This was sized to be a snug fit over the compressor cover, with a section of the ring cut out to allow the compressor discharge nozzle to fit through it. The ring was held in place by longer-than-standard

stainless-steel cap bolts, the ones that also hold the compressor cover in place. Spacers were used to distance the ring from the compressor lugs – without these, the ring would have had to be a very complex shape to provide sufficient 'meat' around the bolt-holes. The wastegate actuator was then bolted to the ring. To make the ring without a lathe would have required extremely careful use of a metal-cutting blade in a power jigsaw and then much filing to shape.

### Suspension bushes

Lathes can be used to turn plastics as well as metals. (In fact, plastics are often easier to work with on a lathe, and much more forgiving of poorly-ground tools, the wrong angles and the wrong feed rates!) One application for which I used the lathe was in machining a polyurethane suspension bush. The commercially available bush wasn't quite the right shape, and so I machined the inner flange to make the shoulder squarer.

Another application was in making plastic suspension bushes from scratch. I used High Density Polypropylene, but they could equally well have been made from nylon or PTFE – both the latter are widely recommended for bearing surfaces in this sort of application. The bushes were easy and quick to make – say 15 minutes each.

### Headlight surround

It's not an automotive application, but it might just as well have been. The aluminium rim being used on a pedal recumbent trike headlight was turned-up on the lathe from – wait for it – the bottom of an old aluminium fire extinguisher! The inner diameter of the fire extinguisher (75mm (3in)) already matched the outer diameter of the stainless-steel drinking cup being used to form the main

The front flange for this custom LED light was machined from the base of an old aluminium fire extinguisher using my lathe.

body of the light, so it was just a case of parting-off the bottom of the extinguisher, and then using the lathe tool to cut out the centre of the base. The rolled edge was already present (to hold the pressure in the extinguisher) and so the whole job took maybe 20 minutes! Without a lathe, realistically it would have been impossible.

### Shortening a nut

Using the lathe, a nyloc nut was shortened so that it provided the right threaded length for an oxygen sensor mount on an exhaust pipe. The section containing the plastic was completely removed, and the nut halved in length. The modification could have been done with a hacksaw and file, but it would have been extremely hard to get it looking as good and would have taken a long time.

The Southbend lathe that did the above jobs has long since gone to another owner, and these days I use a larger, more expensive machine that I bought new. However, I still tend to use the lathe for relatively simple jobs like those described above, and it's not one of my most frequently used machines. But when I need it, it is very useful indeed.

## VERTICAL MILL

Many of the comments I made about the level of expertise required to use a lathe also apply to a vertical mill. To use it with great precision requires skill and training; to use it to reduce the thickness of a piece of steel or aluminium, or to cut a slot in thick stock, is quite easy. You can also use a vertical mill simply as a heavy-duty, rigid drill press.

A vertical mill suitable for home workshop use will have a table around 820 x 240mm (32 x 9½in), and can move around 190mm (7½in) in one direction and 540mm (21in) in the other. Motor power will be around 1500W, and the head will be adjustable for height by rack and pinion. Speed control will be either electronic variation in motor speed or via pulleys and a belt drive, like a drill press. My mill has these dimensions and has a 3MT spindle taper. When buying a mill, also buy a selection of cutters – most look a little like flat-ended drills and can be used for facing or slot cutting.

As with the lathe, it might be illuminating to look at some car jobs I have done with the mill.

### Engine mount

When I added an alternator to an engine that did not originally have one, I needed to make a heavy steel bracket. Part of this bracket bolted to the engine at the engine mount, with the steel of the bracket sandwiched between the engine mount and the engine. The thickness of the steel meant the cast aluminium mount needed to be reduced in thickness by the same amount. This was easily achieved by the mill.

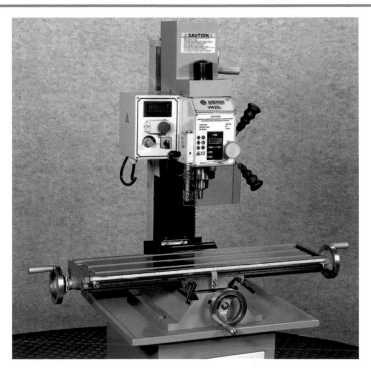

This mill uses a brushless, variable speed drive. As it comes standard with a 3-16mm drill chuck, and has a motor power of 1100W, it is also a powerful drilling machine.
(Courtesy Weiss Machine & Tools)

Here, the thickness of an alloy engine mount is being reduced to give the required clearance for a new alternator bracket.

### Steering arms

When installing air suspension in one of my cars, I needed new steering arms made up from thick steel plate. These arms were to be welded to the new MacPherson struts. I had a local engineering company – a company that normally works on heavy earthmoving machinery – make the arms for me. However, when the arms arrived, I found that the company had been a little generous in their interpretation of my drawing – the arms were massively strong and heavy. I then used the mill to reduce the thickness of these billet steel parts.

If you are doing a lot of fabrication, a mill can be a very useful tool. If you currently have a light duty drill press and are thinking of buying something much more powerful, it might be worthwhile to keep the light duty drill press and rather than buy a heavy-duty drill press, instead buy a vertical mill for both drilling and machining duties.

## AIR COMPRESSOR

As described in the previous chapter, if you want to operate any air tools, you will need a compressor. Most home workshop owners buy compressors that are too small. This means that when spraying paint, or running a bead blasting cabinet, the compressor cannot keep up with the demand and so pressure falls. In the worst case, the operation must cease until the pressure builds up again.

On the other hand, if you are going to be only sporadically operating air tools like a rattle gun, or using the compressor for blowing-out lines or inflating tyres, you will be able to get away with a smaller compressor.

My compressor is a twin cylinder design rated at 190 litres/min (6.7cfm) FAD (free air delivery). It uses a 50-litre (13 US gallon) tank, and has an electric motor power of 2.5kW. It is fine for all my air needs, except for my bead-blasting cabinet – more on that in a moment.

I suggest that you don't buy an air compressor for a home workshop unless you have a specific use that requires air. That is, let the demand for a certain tool drive the need to acquire the compressor, rather than buying a compressor and then finding the tools you can run off it.

Air compressors are potentially dangerous. To prevent the steel tank rusting from the inside out, you must drain the tank of accumulated water periodically. How often you need to do that depends on the humidity of the air and the frequency of compressor use. If a lot of water comes out, drain it more frequently. Because of tank safety, I don't recommend buying a secondhand air compressor. Compressed air itself is also dangerous – be careful when using it to blow out lines and the like, that high-speed particles of dirt don't enter your eyes, and never direct compressed air at anyone.

My air compressor is a relatively recent acquisition in my workshop, but I use it every few days – for operating the

This compressor has a free air delivery of 303 litres/minute (10.7cfm) and draws 16 amps at 240V. Always lean towards a compressor that is larger than you expect to need. (Courtesy Toolstop)

bead blaster, for inflation of tyres and air springs, and for clearing water and dusting-off surfaces.

## HYDRAULIC PRESS

If you've been totting up the cost of the tools covered in this section and your eyes are starting to glaze over, here's a much cheaper tool that is still very useful. A small hydraulic press, permanently mounted on a bench or stand, is a tool with many uses. Mine uses a 6-ton hydraulic bottle jack and is a compact size with a maximum height of 230mm (9in). Hydraulic presses are useful for pressing in (and out) bearings and suspension bushes, and for forming and straightening metal.

Here are three recent car uses I've made of mine.

### Suspension bushes

I recently made a new anti-roll bar for a torsion beam rear axle. These anti-roll bars normally bolt solidly within the torsion beam. However, in this case I wanted the bar to have a little 'give' and so made up new end fittings into which I pressed polyurethane bushes. The hydraulic press made the process simple.

### Clearance depression

I made a new air-cleaner assembly for a car I was turbocharging. The new assembly used a cylindrical air filter that sat within a cylindrical housing made from 150mm (6in) truck exhaust tube (it's pictured in Chapter 7). However, it turned out that this housing fouled the starter motor solenoid. What I needed was a little depression in the housing to clear the solenoid. I filled the tube tightly with sand and capped the ends. I then used the hydraulic press to push a hemispherical former (a tow bar ball) into the tube at the correct point. The sand prevented the whole tube from being compressed into an oval, while the ball made the right-shaped depression at the correct location. Clearance re-established!

### Doming a steel plate

I installed rolling sleeve air springs in the back of one of my cars. Doing so required the use of a steel pipe extension downwards from the original spring seat location on the torsion beam rear axle, and the placing of a plate in the bottom of the pipe extension, on which the air spring would sit. A flat plate doesn't have much strength, so I used the press to form the plate into a dome with a central flat area (on which the spring pedestal sat). This shaping was accomplished by sitting the plate across the open end of a pipe a little smaller in diameter than the plate, and then using the hydraulic press to push downwards in the centre of the plate with a short length of small diameter heavy pipe. The dome was formed surprisingly easily.

A small 6-tonne hydraulic press like this one is a useful and relatively cheap tool to have in a home car workshop. (Courtesy Supercheap Auto)

If you buy a press, be aware that sometimes the metal supports designed to support what is being pressed may be made from cast iron that can break under the load. Replace them with offcuts of thick mild steel bar.

## TUBE BENDER

A tube bender is very useful to have in a home workshop. It would be nice to have a hydraulic exhaust tube bender, but they're very expensive – what I am referring to here is hand bender that can work with small dimension square and round tube. I use a Multi-Bend tube bender, a tool that has been produced since 1989, so a well-proven design. It has a nominal capacity in mild steel of 25 x 2mm (about 1in x 1/16in), and in aluminium of 32mm x 3mm (about

The dome-shaped panel used in the base of this air spring mount was pressed into that shape with my 6-tonne hydraulic bench press.

A square tube steel frame that I bent to shape using a Multi-Bend hand bender. Having a small tube bender is very useful – bent tube is light and strong.

1¼in x ⅛in). Australian-made, the bender works extremely well – especially on square tubing.

The trick in its design is in the internal formers around which the tube is wrapped. The formers are not just simple cylinders, but instead have a curved, raised rib around their centre. When square tube is forced around the former, the raised rib creates a central depression on the inside of the bend. Around the outside of the bend, the outer wall is also depressed – it does this naturally. These indents allow the bend to be formed evenly. The width of the internal former matches the width of the tube, and the square tube is sandwiched between the upper plate (that is held down by a nut and lever) and the baseplate. This prevents the tube spreading during bending.

The bender comes without the lever that you pull on to perform the bend – you provide it. The manual suggests a lever that's 1m (3ft) long; I use the longest that will freely swing within the available workshop space – about 1.5m (5ft). The bender mounts in a vice, or can be bolted to a bench or very sturdy column.

I use mine frequently, with my most recent use being to make a square tube steel frame that supports an air compressor being used as part of a car air suspension system.

## BEAD-BLASTING CABINET

A bead blasting cabinet allows you to blast items that are small enough to fit inside the cabinet. The cabinet comprises a windowed compartment with two thick gloves into which you can insert your hands. Inside the cabinet is a blasting nozzle, and, in the bottom of the cabinet, a reservoir of blasting medium. In use, the item to be blasted is placed in the cabinet, and then, using your hands in the gloves, the blasting nozzle is aimed at the object. The nozzle draws up the blasting medium and fires it at the object being blasted, with the medium then falling to the base of the cabinet to be re-used. The blasting medium might be garnet, soda, glass spheres or crushed glass.

The abrasive effect of the small particles hitting the object at high speed removes material from the surface. For example, an alloy turbo compressor cover, when blasted with soda, can be brought up to as-new appearance in a way no other technique will achieve. Garnet blasting will remove flux from a weld or rust from a tool. I have a bead-blasting cabinet in my home workshop, and it's used much more frequently than I expected.

However, there are some downsides. A bead-blasting cabinet is a tool to buy only if you already have a large air compressor. Even a small cabinet (with a proportionally small blasting nozzle) is a voracious consumer of compressed air. My compressor is rated at 190 litres/min (6.7 cfm) FAD (free air delivery). When used with my medium-sized blasting cabinet, tank pressure rapidly falls as air demand outstrips supply. The cabinet is still quite usable, but if you are blasting large items, you'll need to wait a few minutes for pressure recovery every 30 seconds or so.

A bead-blasting cabinet has a large footprint, and so if you're tight for space, it's not an item to buy. The blasting media also 'wears out' (ie gets its sharp edges knocked off) and so needs to be periodically replaced. If you are doing a lot of blasting, this can be quite expensive. Finally, you need a way of collecting the dust that is formed when the cabinet is being used. Some cabinets have an attachment point for a vacuum cleaner – this should be used, otherwise a lot of dust will exit the cabinet and float around the workshop. Ensure you cannot breathe this dust – either by using a proper mask (not just a paper one, but one that seals to your face and is equipped with full filters) or by powered extraction.

You can also buy small hand-held blasters that can be used without a cabinet. When using one of these blasters, the blasting medium goes absolutely everywhere, so blasting needs to occur only outside and, unlike a cabinet, there's no re-using of the medium. You will also need to wear full top-to-toe protection. I had to repair a cracked chassis rail on a trailer, and I used a handheld blaster to clean that area of the chassis before welding the crack and then welding a plate over the repair. In that case, the hand-held blaster was ideal for the job.

Note that any items that have been blasted must be thoroughly cleaned afterwards, as the dust will have penetrated every crevice. For this reason, you cannot blast items that contain bearings or similar.

I use this blasting cabinet in my workshop. It's a useful tool but it needs a large compressor to run it and it takes up a lot of space. (Courtesy Toolstop)

**Chapter 4**

# Workbenches

A workshop without a bench is a workshop that's of little use. Even in the smallest of workshops, you should have a sturdy, strong and heavy workbench. In a large home workshop, you should aim to have extensive bench space – for example, multiple benches positioned along walls. You'll use the workbench when working with smaller components, when cutting with hand tools, when holding items in a vice, when fabricating – even when overhauling components like gearboxes. In addition, many power tools are often mounted on benches.

So what features does an effective bench need to have?

## HEIGHT

Picking the correct bench height is vital. That's especially the case if you're going to be mounting a vice on it. The top of the vice should be near to the height of your bent elbow, as shown below. That's because when you place an item in the vice and file it, you'll have greatest control if the item is at that height. The same applies for hack-sawing.

Many benches are too low – in fact, the majority are too low. Too low a bench will also force you to bend your back more than necessary. As an example of specifics: I am 1800mm (just under 6 feet) tall and my elbow height is 1120mm (44in). On my working bench, the top of the vice is 1100mm (43in) from the ground. That bench is 900mm (about 35½in) high.

If you are building the bench as a solid surface on which to mount power tools, the required height may change. For example, a drill press needs to be positioned so that the working table is at a comfortable height, even when a small (and hence short) drill bit is being used in it. A grinder needs

Commercially sold workbenches are seldom of the required standard. This workbench uses thin steel angle legs and a lightweight plywood shelf and top. Better to spend the money on materials and make your own. (Courtesy Ryobi)

to be positioned so that the working table is at a height that allows you full hand control of the item being ground. My power tools bench is 850mm (33½in) high. If people of different height are going to be using the workshop (for example, you want a 12 or 13-year-old to be able to use it), you may need to use a lower bench height again.

While often never considered, selecting the correct bench height for the use being made of it is a vital first step in bench decision-making.

## WEIGHT

A bench should be made as heavy as possible. A well-built bench will be heavy as a matter of course (more on construction in a moment), but having a bench that is heavy is something to aim for in itself. Why? A major reason is that when using the vice, you want it to behave as if it's part of planet Earth!

For example, say you have a metal bar in the vice and you want to bend it slightly. If the bench is lightweight, as soon as you pull on the bar, the bench will skate across the floor. Not good! Of course, you can bolt a bench to a concrete or wooden floor – but it simply isn't the same as having a massive weight to work against. The same applies when using a grinder or belt-sander bolted to the bench – the bench needs to be unmoving.

Benches can be made heavier if they are constructed with open shelving onto which heavy items are later placed. The shelves then need to be strong (much stronger than conventional shelves) to support the added weight. For example, you can place heavy portable power tools on

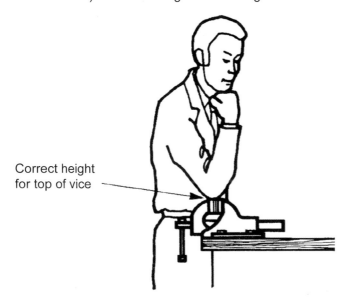

Correct height for top of vice

If you intend mounting a vice on your workbench, the correct height of the bench will depend on your height and the size of the vice you intend using. Test for optimal height in this way.

these shelves (that's what I do) or even load the shelves with bricks or the weights used in weight-lifting exercise machines.

## STRENGTH

All benches should be built super strong. There are three reasons for that.

Firstly – and most simply – you might want to put something pretty heavy on it! That might be a gearbox or even an engine. Any well-built bench should tolerate having 300kg (660lb) or so placed on top.

Secondly, you don't want the bench-top to deflect when you're hammering on it. Whether you mount a small anvil on the bench, you're using a hammer on an item in the vice, or you simply want to centre-punch something sitting on the bench surface – in all cases, the less the bench deflects, the better.

Finally (and this is an important point when you're thinking about building a bench), a strong and well-built bench will last you for the rest of your life. I never really thought about this until I reflected on the bench that I built 25 years ago. I made it when I was a teacher and so had access to the school metal-working facilities (I paid for all materials and consumables). When, one weekend, it was time to weld the bench frame, the MIG welders were low on gas, but I went ahead anyway – and so the welds

An appropriately strong commercially-available work bench. This Norlift Superbench uses a 10-gauge (3.6mm) steel top on a steel pallet frame. Here a 1340kg (3000lb) weight has been placed on top of the bench. (Courtesy Norlift)

weren't nearly as good as they should have been. At the time, I thought: so what, the welds will work well enough. Now, when that bench has followed me to five different home workshops across the country, I sometimes wish I'd waited for the new shielding gas cylinders to arrive. The bench hasn't broken – but I would have taken more care in the construction if I'd realised how long I was going to be working with it!

## SHAPE AND SIZE

A bench can be built in two basic configurations – island or linear.

An island bench sits in the middle of a space – it's accessible from three or even four sides. A bench of this type is rectangular with about 1:1 (ie square) to 2:1 (ie rectangular) length:width dimensions. An island bench has some major advantages. You can work with very wide and/ or long items, because they can overlap the bench edges. For example, you can drill a large sheet of aluminium or plywood. Alternatively, you can mount items at the corners (eg a drill press, bench grinder or anvil) and then easily access all of them. However, because you need to be able to work all around it, an island bench consumes a lot of space – so that space needs to be available in the first

A compact and strong commercially-available workbench made from pallet racking. Lose the castors though, and bolt the bench to the ground or place heavy weights on the lower shelf. (Courtesy Norlift)

A standalone bench that can be moved as required. It is made from 50mm (2in) square tubing, has adjustable height feet and a maple butcher block timber top. The curved cross beam is an elegant touch. (Courtesy Jack Olsen)

## CONSTRUCTION

To give the required strength, a workbench needs to be made of heavy, large-section timber, or thick-wall steel tube or angle. The material needs to be fixed together very sturdily – the timber by using rebates or similar joints, bolted and braced. A steel frame bench needs to be welded or bolted together, usually with gussets or diagonal braces for additional strength. I'll cover building your own workbenches in moment, but at this stage keep in mind that all effective benches need to be constructed in a way that will give them great, enduring strength.

## BUYING A COMMERCIAL WORKBENCH

If you visit any major hardware store, you'll see plenty of commercially-available workbenches. Just pay your money and take one home. The trouble is, most of these are absolutely useless. They are weak, too light, won't be durable and are often the wrong height. They are ideal for hobbies like model building and electronics (but then again, why not just buy an old kitchen table for this?) but for working on cars, they have little to recommend them.

I strongly suggest that you build your own workbenches, but if you're not able to do so, here are some buying tips.

Note that it's best if you can see and touch a workbench you are considering buying, but if you must buy as a flatpack or online, here's a simple test. How heavy is the workbench? Each of the self-built workbenches covered later in this chapter weighs at least 100kg (220lb). If the workbench you're looking at online has lots of promotional lines about 'extra heavy-duty construction' and 'made to last' – and yet weighs only 30kg (66lb), steer clear.

place. I like using an island bench and use one as my main working bench.

A linear bench mounts along a wall. Its length:width ratio is often about 3:1 or 4:1. A linear bench takes up less workshop space. If building a linear bench, don't fall into the trap of making it weaker just because it may partly be supported by the wall – for all the same reasons as an island bench, it needs to be super strong. I use linear benches to support tools like a drill press, grinder and small hydraulic press.

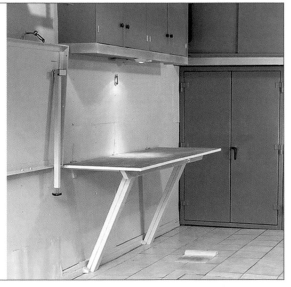

If you're really tight for space, you can use benches that, when not being used, fold up against the wall. The left picture shows two benches in the stored position; in the right-hand picture one of the benches can be seen folded down. This bench is 71cm (28in) wide and 198cm (78in) long. It uses a 19mm (¾in) plywood top, with two layers used over most of the bench top. (Both pics courtesy Jack Olsen)

So, what are the key points to look for if buying a workbench?

Firstly, is the bench an appropriate height for you? If it is too low, can it be easily raised – for example by bolting or welding additional supports to the legs? If it is too high, is lowering a straightforward job?

Can you wriggle the bench – try to push it sideways and back and forth. If the bench immediately flexes, imagine what it will do with a piece of steel clamped to it – a piece of steel that you're trying to bend! If the bench is sturdy but slides along the floor, can large weights be added to it? For example, is there a shelf beneath the bench that is strong enough to take 100kg (220lb) of added weight? Can the bench be easily bolted to the floor?

Is the bench top stiff? The material from which the bench top is formed needs to be thick and not springy, so that when you hammer on it, the material remains 'dead.' If it springs back, then working on it will be difficult. The best commercial benches use thick timber or composite board. A thin steel top with nothing underneath will be useless – beware benches with thin stainless-steel tops that look great in the shop!

Does the edge of the bench top overhang sufficiently that G-clamps (sometimes called C-clamps) can be used to clamp items to the bench? If the edge of the bench is flush with the supports, it will be hard to clamp items to the bench. Furthermore, what will the clamps hold on to under the bench top? If the surface under the overhanging edge of the bench is cluttered with brackets or braces, it will make it more difficult to use the clamps.

Will it be straightforward to bolt a vice to the bench? Again, when you look underneath, some benches have supports and brackets that are placed in the way of the bolts you'd use to mount a vice.

If you can find a bench that matches these criteria and is less than your available funds, buy it. However, chances are that you won't be able to find a bench that's really good, so you'll need to build your own. That can be easy and cheap, will give you a bench that will last and be good to work on. In the sections below I cover three different benches that I have made over the years. All are still in daily use.

## MAKING A SIMPLE STEEL AND TIMBER WORKBENCH

To reduce cost, it makes most sense to find some cheap materials, and then develop the workbench design from there. In fact, the incentive to build the bench I am covering in this section came when I went to a local garage sale. On sale were four lengths of 51mm (around 2in) square steel tube with a hefty 3mm (⅛in) wall thickness. Each had a 'foot' welded on it – I think they'd previously done duty as veranda posts. In addition, there were two 3m (10ft) lengths of heavy 242 x 46mm (9.5 x 1.8in) timber planks. These had

A well-equipped bench suitable for smaller work. It uses a solid oak top and the supports are made from cut-down cabinets. Note the false upper drawer on the right that houses power outlets and an extendable air hose. (Courtesy Jack Olsen)

perhaps been used as roofing rafters or something similar. These items made up nearly all the materials I needed to build the bench – and at garage sale prices!

One of the beauties of building your own bench is that you can tailor the design to suit the materials you have available – for example, if the steel and timber are thinner than desirable, you can build in more braces. In fact, thinking I'd probably need some braces, I sourced four

An island workbench. This workbench, that I first built about 25 years ago, has followed me to five different home workshops across the country. It uses an angle steel frame supporting a thick wooden top with a working surface of replaceable Masonite hardboard. Extra weight and stability are provided by storing heavy power tools on the internal shelving. I do nearly all my work on this bench.

This workbench uses a frame made from heavy-gauge square steel tube. A thick timber top is glued to the frame with industrial strength adhesive. Note the frame bracing – one diagonal at the rear and two cross-braces at each end. This is a sturdy and effective workbench that was made from materials mostly bought cheaply at a garage sale.

short lengths of 17mm (a bit under ¾in) diameter steel bars (rebar) normally used for reinforcing concrete. This steel was bought from a local salvage yard.

The top working surface of a bench can be stainless steel sheet, or, my preference, a thin layer of Masonite. (Masonite is a high-density pressed wood sheet.) Masonite is replaceable after it wears, doesn't easily dent and can be kept clean with a wipe-over. It also absorbs spills. So I added some 5mm (a little under ¼in) thick Masonite sheet – this was the only material bought new. With that, the list of construction materials was complete.

The bare minimum in tools for building a bench of this type are a hacksaw and electric drill. However, a welder will make things much stronger. A friction cut-off saw will also make cutting up the steel much easier, and a table saw will make cutting the timber to size neater and quicker.

The first step was to decide on the bench height and width. With a strictly finite length of square tube available, the bench length dimension was based mostly on what was left over after cutting the uprights and width pieces to size. For the framework, I used an upright length of 840mm (33in), and a width of 580mm (22.8in). (Note: the actual height and width of the bench are greater than these dimensions; the height is increased by the thickness of the top timber and the width is increased as the timberwork overhangs the frame a little.) A frame length of 1720mm (68in) was used.

The framework was in a traditional 'table' form – four

pieces of tube welded to form a rectangular frame and then the legs welded on. Halfway along the top I welded another piece of square tube across the width.

Such a frame is already pretty strong – primarily because of the thick walls and reasonably large tube dimensions. But what if a force is applied end-on to the frame? In that case, the rectangular frame will distort into a rhomboid shape. To prevent this distortion, a diagonal cross-brace was placed across the back of the bench frame. In one direction of distortion, the cross-brace will be in compression, and in the other direction of distortion, extension. Therefore, if you're going to use only one cross-brace, it needs to be made of material rigid in compression as well as extension. This was achieved in the case of my bench by making the cross-brace from an additional piece of heavy-wall square tube.

The same potential distortion could also occur if a force is pushing backwards on the bench. Another heavy-duty cross-brace could have been used at each end of the frame – but I'd run out of the square tube. Instead I used the round concrete reinforcement rod I'd bought earlier. However, this rod is not very strong in compression – it's much stronger in extension. The rods were therefore used as dual cross-braces – so whichever the direction of distortion, at least one rod will be in tension. After welding the frame together (you could also bolt it), it was painted with epoxy enamel.

The heavy timber planks were then cut to size and glued to the frame. Modern glues are incredibly strong, cheap and easy to use. I used three cartridges to glue the planks

The bench equipped with a hydraulic press, grinder, small drill press and small anvil. This set-up is ideal as a compact ancillary bench, and can be mounted against a wall without taking up much space. If even greater rigidity is needed, it can be bolted to the floor.

to the frame and the planks to each other. When the bench is in hard use (eg something is being hammered in a vice), most of the forces acting on the glue will be sheer forces, something the glue is very well able to withstand. The glue also acted as a filler where the planks weren't dead flat.

After the glue had set overnight, I ran an electric plane over the surface to true-up the planks (a normal hand plane could have been used – or you could have used timber that was unwarped!) and then the Masonite top surface was cut to size and held in place with small brads that were recessed a little below the surface. The Masonite could have been glued into place, but that would have made it hard to remove and replace – and one of the advantages of this material is that it can be easily and cheaply replaced after it wears.

This bench was made primarily to support some tools, rather than act as a dedicated working surface. On the bench were placed an 8in electric grinder, a hydraulic press, a drill press, and a small anvil. Each was bolted into place. The bench is easily strong enough, rigid enough and heavy enough to support the anvil – and it would also be fine if a vice later needed to be attached.

## MAKING A LINEAR TIMBER BENCH

When I was setting up a new shed-based home workshop, the need arose for a long workbench that would run along a wall.

As with the bench described above, the availability of the construction materials dictated the design. In this case, knocking down an old garage had given me a plentiful supply of 290 x 50mm (about 11 x 2in) hardwood planks – the old roof rafters. These would form the underpinnings of the bench top. In addition, there was an array of other heavy timbers that could be used for the supporting legs.

The first step was to lay out the long, thick planks that would form the bench top. Using supporting crates and a small step ladder, I placed the planks at a height that I thought would be effective. In this case that was 870mm (just over 34in). Note that this is a little lower than the 900mm (35½in) height that I earlier nominated as best suiting me. There were two reasons for selecting the lower height. Firstly, this height better matched the existing power outlets and exposed wall frame member, and secondly, my then 8-year-old son was also going to be using this bench. At this stage, I also assessed the bench width that I wanted, settling on 590mm (around 23in). The bench was six metres long – that's nearly 20 feet!

With the planks positioned at the correct height, I could then start building the supporting legs. I used 70 x 50mm (about 3 x 2in) timber to form rectangular supporting frames, each rebated on the uprights. The joins between the uprights and the cross-pieces were screwed together. To give greater stiffness, I placed a 75 x 25mm (3 x 1in)

A long linear bench, made from salvaged timber and built into place. Note how the bench-top wraps around the shed upright to give longitudinal stiffness. The bench-top is positioned just under the existing power outlets, and cross-braces have been used on each of the supporting frames. This bench is all timber – you can still make a very strong bench without needing a welding machine.

hardwood diagonal on each frame, again screwing them into place.

The frames were made on the flat concrete floor and then inserted under the supported bench top, being located just under 1m (3ft) apart. As the frames were placed into position, they were attached to the bench top by long screws inserted from the top. In other words, the bench was constructed in situ – it was much too heavy to build it elsewhere and then move it into place.

The supporting frames were then dyna-bolted to the concrete floor, using two 12mm (½in) bolts on each frame. A rented hammer drill was used to drill the holes in the concrete. The floor spaces between the frames remained completely clear – this allowed boxes or crates to be easily slid under the bench-top for storage. Lengthways movement of the bench was resisted by the frame of the shed, around which the bench-top fitted very snugly (a light hammer push fit, in fact).

As with the bench described previously, Masonite hardboard was used to form the bench top. The bench top was oiled, using new unwanted engine oil that I had. The oil was applied with a brush, left for 15 minutes to sink in, then the surplus was wiped off with a cloth. The resulting surface was dry to the touch, didn't show oil stains, and could be easily touched-up should it be scratched. (Don't apply used engine oil – it can contain nasties.)

The bench was sturdy enough for tools like vices and grinders to be bolted straight on. However, note that if the bench were to be used for really heavy hammering,

**A bench made using second-hand pallet racking to form the heavy-duty steel frame. This bench was made specifically to mount power tools – the hand shears shown here are for display purposes only!**

additional longitudinal rail supports beneath the top would be needed to stiffen it.

## MAKING A PALLET-RACKING POWER TOOL WORKBENCH

These days, bench-mounted power tools like grinders and sanders are incredibly cheap. That's great – but where do you mount them? Ironically, while stands and supports are available for this type of tool, the cost of the stand is often as high as that of the power tool itself! With that in mind, I decided to build a workbench to specifically suit some of my bench-mounted power tools.

Note that, to build a similar bench, you'll need access to a welder, angle grinder and (preferably) a friction cut-off saw or horizontal metal-cutting bandsaw.

The first point to consider when designing a steel-frame workbench of this type is to understand the vast length of steel that's needed. The bench constructed here used something like 16 linear metres (around 52ft) of steel! Pay even a low amount per length and the total adds up alarmingly quickly. The trick is to find steel that no-one wants – and one answer can be found in pallet racking. The beams used in pallet racking are available secondhand very cheaply – that's especially the case if they comprise odd collections of different cross-sectional sizes and lengths.

The pallet racking beams used here were left over after I bought a heap of secondhand beams for racking in

my home workshop. I bought 50 beams secondhand and used about 36 of these in the workshop racking. Of the 14 remaining beams, I used 11 to build the bench shown here. The final cost of the steel was ridiculously low.

Despite the fact that this bench is not going to be hammered on, or carry really big weights, I chose the same approach as described above for the bench top – thick timber with a thin sheet of Masonite on top. Why change something that works so well?

The first step in the construction was to cut the pallet racking beams to size. I used a friction cut-off saw to do this. It could also have been done with a hacksaw (a nightmare) or a cutting disc in an angle grinder. I took the friction saw outside, so the huge amount of generated swarf didn't end up all over the floor of my workshop.

The beams that would form the top of the frame were then arranged on the workshop floor and tacked into place. It's very important that everything is square – both in plan view and elevation. Once tacked into position and checked for squareness, the beams could be fully welded together. The rest of the bench frame was then cut, assembled and welded – the photos show the design that was used. Note the large number of supports that have been used for the lower shelf – on this shelf will be placed heavy objects to stabilise the bench. The additional supports will stop the shelf surface from sagging under this load.

After the frame was completed, it was given a coat of

**The frame of the bench, showing the pallet racking rectangular tube welded into a strong structure. The lower shelf has plenty of cross-pieces to give good support to the shelf, allowing thin material to be used for this shelf. Placing heavy items on this shelf stabilises the relatively long and narrow bench.**

paint. Note that if the pallet racking beams are powder-coated (rather than painted), the surface should be heavily scuffed with sandpaper to allow the paint to key into the surface.

The next step was to mount the timber top. A few different approaches can be taken. Simplest is to use building adhesive to glue the pieces of timber to the metal frame – as was done for the first bench covered above. Another approach is to drill countersunk holes and use bolts or self-tapping screws to firmly attach the timber planks to the frame – that's what was done in this case. Either way, when placing the planks, use adhesive between the planks so the top becomes a homogenous whole.

If the planks form an uneven top surface, you can run an electric plane or sander over them. If you don't have either of those, you can use a hand plane or, as a last resort, a hand saw used at an angle of about 5 degrees – and so 'rasping' the surface rather than cutting it. With the timber flat, the Masonite top surface was installed.

The heavy timber top, with gaps filled. This uneven surface was levelled using a hand saw angled at about 5 degrees to work as a heavy-duty rasp. Masonite hardboard then formed the upper working surface.

A neat touch is given to benches by the use of Masonite hardboard on the top and front surfaces, with the leading edge rounded by a plane or sander. The hardboard can be oiled with new (but unwanted) engine oil to give a hard-wearing finish.

The lower shelf surface could then be placed into position. Any composite board can be used – because it is well supported, even relatively thin material won't unduly droop. The board can be glued and/or screwed into place. I used the coated particle board from flat pack bookshelves that had not survived a house move.

## COMMON DESIGN POINTS OF WORKBENCHES

You can see that in each of these self-built designs, several themes show through. The benches are:

- massively strong
- have a stiff but 'dead' top surface comprising thick timber topped with thin replaceable hardboard sheet (for jobs like gearbox or engine disassembly, a stainless-steel sheet on top would also be fine)
- can be either bolted to the floor, or have shelves on which heavy weights can be placed, so stabilising the design
- use salvaged and/or secondhand materials

The resulting benches are effective, long-lasting and low in cost.

**Chapter 5**

# Storage

I must confess it has taken me literally decades to realise how important it is to have an orderly home workshop. Not necessarily neat and tidy, mind you, but one where you can put your hands on what you want in the least possible time. When you want a 12mm (½in) deep-style socket, you need to be able to go right over and get it. When you want a 6mm flanged nut and bolt and a mudguard (penny) washer to suit it, again they need to be available in moments.

I think that lots of people become disenchanted with working in home workshops when they can never lay their hands on what they want – and so every project takes an eternity. And one of the main ways of achieving an ordered workshop is to have top quality storage. That in turn means there's a place for everything – and everything can be in its place.

In this chapter, I want to cover effective storage – storage of car parts, of tools, of fasteners, of tube and bar stock, and so on.

## TOOL STORAGE

I choose to use three approaches to storing my tools.

For those tools that are used most frequently – my main tool kit – I use a traditional 'rollcab' tool chest. This is a large steel unit that has drawers mounted on ball-bearing extension runners. In this tool chest I have my socket set, spanners (wrenches), pliers and so on. In the deep bottom drawer, I place my most-used power tools. The rollcab is on large castors and so can be wheeled about as required. The rollcab keeps these tools sorted and allows quick and easy access. The downside is that this is an expensive approach to storing tools. Also, be aware that you should never have multiple drawers open at once – even an apparently stable tool chest can overturn if more than one drawer is open.

The tools I use much less frequently are stored in large open-top trays and boxes located in heavy-duty cupboards positioned against the workshop walls. For example, one of the trays contains bolt cutters, spring compressors, heavy

I use a 'roll-cab' storage chest like this for my most frequently used tools. It places the tools conveniently to hand and allows easy organisation. However, tool storage like this is expensive. (Courtesy Toolstop)

I hang tools on the inside of heavy duty cabinets that I made. The doors are sufficiently strong to support tools like the files shown here.

If you're equipping a home workshop from scratch, storage cabinets from the one manufacturer look good and work well. (Courtesy Car Lifts Plus)

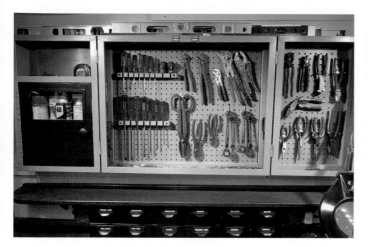

Pegboards with hooks are an easy and versatile way of storing tools. Here, a two-door cupboard is shown in its open position. (Courtesy Jack Olsen)

Small components such as switches, terminals and plugs are best kept in drawers such as these.

battery cable crimpers, pry bars, and so on. These tools aren't precisely organised like the ones in the toolcab – instead, they're just piled into the trays. I know which tools are in which trays, but when I need a specific one, I have to spend a few moments digging through the tray. The benefit of this approach is that storage is cheap, and a lot of tools can be fitted in a compact space.

Finally, I hang certain tools on dedicated hooks and supports. In my workshop, I use the inside of cabinet doors to do this. The cabinet doors carry files, hammers and adjustable spanners. (More on constructing these cabinets in a moment.)

When devising a storage system, keep the following aspects in mind.

* The most frequently used tools need to be quickly accessible and best organised.
* Some tools (eg files) should not be stored in contact with one another.
* Expensive and elaborate tool storage might look good, but it's better to spend that money on the tools themselves.
* Use labels on drawers so that you can more readily find seldom-used tools (and it's easier for others who may be helping you).

## STORAGE OF NUTS AND BOLTS

In many home workshops, fasteners like nuts, bolts and washers are often stored in small parts drawers, made of either plastic or steel. However, I typically find these too small – they're fine if you have only a couple of bolts and nuts in one size, but if you have 50 or so large bolts, they no longer fit. (However, I find these drawers perfect for storing very small items like electronic components, split pins and circlips.)

For fastener storage, I prefer to make my own drawer system. The drawers are formed from kitchen baking trays working in a timber frame (see pictures on page 64). Rather than the base of the pan sliding back and forth on the timber, it's the flanged upper edges of the trays that slide on runners.

I keep a look out for shops selling steel baking trays and cake baking pans, and buy them only when they're on super special offer. When I find trays that are relatively deep and torsionally stiff, I get 30 or 40 of them. When buying lots of trays, saving even a little on each quickly adds up, so check pricing before spending the money. Ensure that the trays are coated with non-stick material – this stops them rusting, and lets the drawers slide easily on the runners.

The trays used in my current main fastener storage area are each 280 x 180 x 60mm (about 11 x 7 x 2.5in), and 36 of them are used in a 6 x 6 array. In my design, the end-pieces are made from sheets of laminated pine, so giving the required 280mm (11in) depth (thick plywood would be cheaper). The runners on which the drawers slide were made from 60 x 20mm pine (about 2 x 1in). You need a lot of this timber (something like 24m (72ft) for my 36-drawer design). It's the width of the timber that sets the vertical spacing between the drawers, so ensure it's wide enough to provide this clearance. The photographs show the design, which is self-explanatory. With the timber cut to size, it's just a case of lots of gluing and screwing. Use a temporary spacer to set the gap between the adjoining runners – then it's easy to keep the runners parallel.

These drawers are made from coated baking trays, mounted in a wooden frame. The trays slide on their upper flanges and can be easily removed if necessary. There are 36 drawers holding a vast number of fasteners.

This photo shows the approach to building the frame that supports the baking tray drawers. Ensure you keep the slots parallel and matching left/right. Gluing and screwing gives a durable assembly.

The drawers hold a massive number of fasteners, with the fasteners easily visible for quick access. Where I want some specific fasteners located near the job I am working on, I take out the drawer – it slides straight out the front.

## SHELVES THAT ATTACH TO THE WORKSHOP WALLS

If you have a workshop with strong walls, you can easily add heavy-duty shelving to them. The trick is to make the shelves strong enough to support large loads over a long time without sagging. One good approach to achieving this is to use a steel frame that has both vertical and horizontal supports.

My home workshop is constructed within a shed that has exposed internal steel framing. I decided to build some shelves that would use shelf frame verticals supported by the shed's wall girts (the frame longitudinals), with the verticals of the shelf frame being tek-screwed top and bottom to the girts. This allows the structure of the shed to support the shelves, saves on material and works well.

The first step was to cut some medium gauge 40mm (about 1.5in) square steel tube to length. Five verticals

Picking a single storage container size and then optimising the storage for it allows you to really fill the available space. There are 21 containers here. (Courtesy Jack Olsen)

The shelves in place, supported by the frame of the shed. The shelves themselves are made from thick and strong floor panels, cut to size.

and 15 horizontal shelf supports were needed. To speed up this cutting, a friction cut-off saw was used with a stop that provided the right length of material. A cut was made, the tube slid through the saw until it hit the stop, and then another cut was made. Compared with individually measuring each piece, this speeded-up the process by a factor of about ten! Next, a simple welding jig was made from scrap timber. This located the vertical and the three horizontal shelf supports, placing the members in the right locations and ensuring that each bracket assembly was square.

The use of the 'production line' approach meant that all the brackets were identical, and were finished quickly. The vertical spacing between the shelf supports is 260mm (about 10in) and the supports project out the full depth of the shelves – again 260mm.

The shelves themselves were made from water-resistant particle board flooring. This was bought in full sheets in a thickness of 19mm (¾in). Don't confuse normal chipboard with proper structural flooring panels; the flooring material is much more durable (especially if there is moisture in the air), and is also stiffer over the long-term. It can be cut with a table saw or portable power circular saw.

The approach of using the shed walls for support, welded steel square tube for the frame, and shelves made from thick, water-resistant flooring panel has resulted in a strong and durable shelf system.

## STORAGE FOR SHORT LENGTHS OF BAR AND TUBE

If you have lots of pipe or rod offcuts in stock, it can be difficult to work out how to store them. You want them all to be accessible – and yet still not take up much room. Piling them in a heap is a poor approach – you can never be sure what's down the bottom of the pile!

One approach is to make a rack that stores the offcuts vertically, and off the floor. A box-like steel frame is welded-up and then steel mesh is welded into place top and bottom. This mesh can be salvaged from offcuts of concrete reinforcing mesh or the like – it needs to have large openings in it and be strong. Suspended a little below the bottom mesh is a solid panel – I used some more of the particle board flooring material that was used in the previously described shelves.

And how does it work? The tube and bar stock are slid through the top mesh panel downwards until the lower ends are trapped in the bottom mesh. This vertical storage allows you to easily see what stock is available. Keeping it off the floor allows you to store other materials beneath it, but still be able to reach the offcuts, and the mesh keeps the different types of stock sorted. Note that if mounting the storage assembly in the air, make sure that it is well supported – it can get very heavy if lots of material is placed in it.

This holder is excellent for storing short offcuts of pipe, bar and tube. In operation, the bar or tube is slid down through the top mesh until its base is trapped by the lower mesh. Note also the storage space beneath the holder, that is mounted well off the floor.

Heavy duty steel cupboards with no less than 139 parts bins located inside. Before using a configuration like this, ensure the doors are strong enough and don't forget to recess the inner shelves sufficiently to provide clearance with the doors shut.
(Courtesy Jack Olsen)

65

To make some very heavy-duty cabinets, I started with a simple jig to hold the square tube in place while it was being welded (left picture). This side frame is 2.1m (about 7ft) long and, including the base, has five shelves.

Right: the jig for the shelf supports. When making multiples of anything, a jig speeds up the process.

Below: the steel frame of the cabinet. Note that the front of the shelves (shown face up) have been recessed by 100mm (4in). This will allow room for tools to be hung on the inside of the doors.

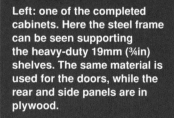

Left: one of the completed cabinets. Here the steel frame can be seen supporting the heavy-duty 19mm (¾in) shelves. The same material is used for the doors, while the rear and side panels are in plywood.

Right: View inside one of the completed cabinets. Note the shelves and removable plywood open-top boxes that sit on them. As the ones at bottom-left show, these doors and shelves are strong enough to support good loads like these bolts.

## HEAVY-DUTY CABINETS

When I was reorganising my workshop recently, I looked around for some heavy-duty steel cabinets. I wanted them to be about 2.1m (just under 7ft) high, about 1 metre (3ft) wide and about 50cm (20in) deep. I was thinking of buying a five-shelf design, and an important aspect was that I wanted to be able to place up to 30kg (66lb) on each of those shelves.

I looked and looked for suitable cabinets, finally settling on some kit units from a hardware chain. These had a claimed shelf weight rating that matched my requirements. I bought two of the kits and got them home, only to find that they were meant to be assembled with tiny self-tapping screws. I assembled one kit, drilling holes and using much stronger bolts and nyloc nuts, rather than using the self-tappers. The kit went together all right, but the stiffening of the structure that you expect as assembly nears completion never came. The finished cabinet had terrible torsional strength, the shelves drooped massively with loads anything remotely near 30kg (66lb) per shelf, and the whole thing struck me as dangerously inadequate. I took the other cabinet back to the shop and got a refund.

So what to do? I decided to build three of my own-design cabinets. They use a very strong steel frame; floor-grade 19mm (¾in) particle board for the shelves, front doors and tops; and have an innovative design that allows a lot of tools to be hung on the inside of the doors.

The frame of each cabinet comprises about 30m (just under 100ft) of 30 x 30mm (about 1¼in) square steel tube. Tube wall thickness is 2mm (just over 1⁄16in). Thirty metres of tubing per cabinet sounds like an enormous amount of steel, but that's how much it takes to give the required support and stiffness. The frame was welded in the shape of an elongated vertical box, with a full perimeter frame provided for each shelf. In addition, each shelf (and the top and bottom) have a single cross-brace support. The shelves are recessed 100mm (4in) rearwards from the front. This gives room for tools to be hung on the inside of the doors. Each cabinet has about 2m² (21ft²) of hanging tool storage inside the doors.

Each cabinet has two doors, and these doors each use three hinges. The hinges are welded to the steel frame and screwed to the doors. As mentioned, the doors are made from floor-grade 19mm (¾in) particle board; this is sufficiently stiff to support the weight of hanging tools – even heavy hammers or clamps. Side and rear panels are made from light duty 7mm (about ¼in) plywood. You could easily leave off the side and back panels, especially if the cabinets are against a wall and located side by side – the plywood panels were a surprisingly expensive part of the construction. For the doors, I used magnetic catches top and bottom, and made my own simple handles. Overall, the cost of the materials was about 25 per cent more than the flimsy kit cupboards.

The resulting cabinets are very strong indeed (able to take 100kg (220lb) – *per shelf*), and the doors can support about 25kg (55lb) each themselves. It was wonderful loading up a cabinet and then finding that the doors shut just as precisely as when the cabinet was empty! If you are after really heavy-duty designs, I suggest that you look carefully at making your own shelves and cabinets – they'll then have the required weight carrying capacity, fit precisely into the available space, and save you money.

## BOXES AND TRAYS

Over the years, I've bought numerous plastic boxes and trays to store parts and components, but with very few exceptions, subsequently they've always become brittle and have broken. The two exceptions are crates that are designed to hold drink bottles, and the parts drawers that clip into racks. Both of these container types are made

**In my workshop I make extensive use of storage trays like this. The base of each tray is made from 19mm (¾ inch) chipboard flooring with the side of the trays made from 10mm plywood. Glue and screw the assembly together and the trays are strong enough to even store offcuts of steel plate and bar, as shown here.**

from polypropylene – to identify containers made from polypropylene, look for '5' in the triangular recycling symbol moulded into the item. Polypropylene has good oil and solvent resistance, and is durable over the longer term. I use milk crates extensively for storing car components like suspension springs, instrument panels (wrapped in rags), trim panels and the like. If you intend to do the same, organise your shelving so these crates are a neat fit – it will save a lot of space.

However, most of my open-top boxes and trays are constructed not from plastic but from plywood. I make these myself – I've now made dozens and have found them to be strong, light, cheap and durable.

Smaller trays use 9mm (about ⅜in) thick plywood for the sides and bottom. The trays are about 370 x 100 x 110mm (about 14 x 4 x 4½in) and are glued and nailed together. White PVA woodworking glue and small brads results in trays that are surprisingly stiff and strong. Medium-size trays are made from thicker 12mm (about ½in) ply, again used for both the sides and bottom. PVA glue and screws or nails can be used to hold them together. Finally, I make large trays that use 19mm (¾in) particle board flooring for the base and 12mm (½in) ply for the sides. I glue and screw (not nail) these together. A typical size for a large tray is 820 x 460 x 180mm (about 32 x 18 x 7in). If you live in a very humid environment, use marine plywood – more expensive, but it will last forever.

Each of the trays described above is strong enough to be filled with nuts and bolts, brass fittings or the like. The larger trays I use to hold my assorted nuts and bolts (I never throw away a good nut or bolt) and the open tray design allows the easy locating of just the right nut or bolt that's wanted for an unusual use. When filled with bolts, a medium-sized tray of the sort described above, weighs 25kg (55lb) – these trays are *strong* in a way no plastic box of a similar size really is.

If you want to make your own boxes or trays, but don't have wood-cutting facilities, get the shop to cut the ply into long strips of the required width. Cutting the long strips to the correct lengths is then easily done with a handsaw.

### STORING UP HIGH

If you have a high home workshop, don't overlook the large amount of space you have to store items above the working area. I use three approaches. If the items are light, large and unwieldy I hoist them skywards and use chain or ropes to hang them from rafters. Bicycle lifts are good for doing this. If the items are smaller in size, I hoist a platform (eg an old farm gate) upwards, and use it as an elevated high shelf. This approach works particularly well when you have a roof height of around 4m (about 13ft) or above. If you have a really high roof, you can use tall pallet racking, or for storing items like reels of cable, a vertical column with

A car exhaust stored by simply chaining it to the roof.

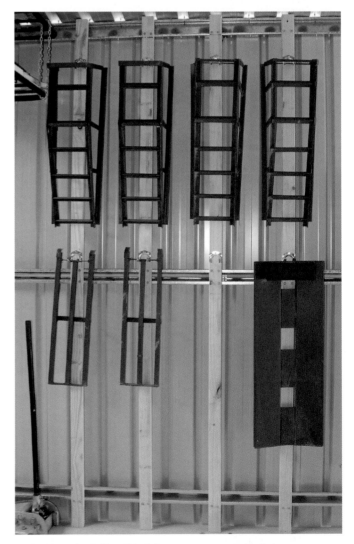

Using the full wall height to store ramps and a creeper.

projections on which you can hang the cable reels. Always secure high-mounted items really well – I typically double-up on the fastenings, just to be sure.

**Extra storage located above the entrance doors, utilising an old farm gate and chains.**

There is extensive storage space in this workshop, but it's been integrated so well it needs a second look to see it.
(Courtesy Jack Olsen)

Chapter 6

# Ramps, hoists, and pits

Whether you're changing a wheel or changing an engine, working under the car is a common occurrence in a home workshop. And if you're under the car, you must have lifted it some way. In this chapter, I want to look at a range of ways in which you can lift a car to allow you to work under it – or alternatively via a pit, lower the ground under the car! I'll also cover engine cranes – where you lift part of the car, rather than all of it.

## JACKS

A jack is mechanical means of lifting a car. Nearly all cars come equipped with simple jacks (invariably of the screw design) that are used to lift a car in order to change a flat tyre. With very few exceptions, these jacks are as cheaply made as the car manufacturer (or their supplier) could achieve. Standard car jacks are unstable on all but the firmest, flattest surface, are not rated for very high mass, and are lots of effort to raise and lower.

Do I have any standard car jacks in my workshop? The answer is actually 'plenty,' but they're invariably not used for lifting the car (instead they're used to compress a suspension arm, or wedge apart a steel frame that has closed-up after welding – things like that).

Forget the standard car jack – in a home workshop you need a good general purpose hydraulic jack. These fall into two categories – bottle jacks and trolley jacks. Bottle jacks are compact jacks that have a small base and an extending, multipart post. In use, they have a lot of the deficiencies of the standard car jack. I have a bottle jack, but I rarely use it except for specialised purposes. A much better jack is a trolley jack. A trolley jack, as the name suggests, has wheels. It has a lever arm that lifts by the action of a hydraulic cylinder. Trolley jacks are stable and can lift high – both excellent attributes. It also helps if the high-lifting, stable trolley jack has a low initial height, allowing it to be slid under a low car.

When selecting a trolley jack, look for:
* appropriate weight rating (eg twice the weight you expect to ever lift)
* a high maximum raised position (eg 500mm or 20in)
* a low minimum height (100mm or 4in)
* large wheels
* appropriate standards ratings

Some people choose to go for lightweight jacks (eg those made of aluminium) but you will pay a lot more, and, in typical home workshop use, the lower weight makes little difference. However, what *does* make a difference is a 'fast lift' facility. These are sometimes achieved by a separate foot operation of the jack – you use this to lift the jack until the load is being borne, and then use the handle to lift the car. In the case of lightweight cars, the foot pedal may be able to be used to lift the car as well. Note that jacks need to be maintained – the pivots points greased (look for

Buy the very best hydraulic trolley jack you can afford. This one is rated at 3 tonnes, extends to 570mm (22½in), and yet still fits under a car that's only 100mm (4in) above the ground. (Courtesy Toolstop)

the provision of grease nipples when buying) and the oil periodically changed and air bled from the hydraulic circuit.

A good quality, easy to use, safe and effective jack is one of the first purchases you should make for your home workshop. I suggest that you buy a jack a little better than you think you'll need – you'll never regret it.

## JACK-STANDS

A jack is used to lift the car – not to hold the car up while you are working underneath it. To support the car, place jack-stands under the elevated car and then lower the jack until the weight of the car is supported by the jack-stands. Jack-stands are available quite cheaply and, if they have the appropriate standards markings and weight ratings, even low-cost jack-stands are fine to use. Jack-stands are adjustable in height, with pin-through-hole being simple and effective. Other stands use a ratchet design, but I've never trusted these as much as a pin in simple double-shear!

At minimum, you need four jack-stands (one for each corner), and having more than four is recommended. I always use more jack-stands than is actually needed. For example, if I have the front of the car in the air, I will use two jack-stands under the front jacking points, and then another two positioned at about the mid-point of the car. The second two are not normally touching the car; they're positioned to catch the car should it fall off the other two jack-stands!

Never work under a car supported only on a jack. Instead, use jack-stands like these to support the weight of the car. (Courtesy Toolstop)

General-purpose ramps are excellent for quickly raising the car to perform routine service like oil changes.

Having extra jack-stands is also useful if you have a spare car that is being stripped for parts, or a car that is having bodywork performed on it. At the moment, I think in my home workshop I have 12 jack-stands – perhaps overkill, but I never run out!

When buying jack-stands, look for:
* appropriate weight rating (eg twice the weight you expect to ever support)
* a high maximum raised position (eg 500mm or 20in – and make sure that this matches the maximum height of the hydraulic trolley jack you've bought)
* wide, stable base
* secure locking mechanism
* appropriate standards ratings

## RAMPS
Ramps vary from short and steep to long and shallow in angle, from those that lift only the wheels at one end of the car to those that lift the whole car.

The most common ramps are made from angle steel and lift just one end of the car. These are often used when changing the engine oil, for example. The ramps are positioned ahead of the front wheels, then the car is driven up the ramps. The oil and filter are changed, then the car is reversed down the ramps. Because they are made in large numbers, these ramps are cheap and readily available. Ramps of this type typically lift the car by 220mm (about 9in). For car maintenance jobs, that height is sufficient, and the ease with which the car can be lifted makes this type of ramp very convenient.

However, for uses other than routine car servicing, these ramps do have some negatives. First, the wheels are not in

the air – they're still supporting the car. You cannot therefore do any brake work or suspension work, for example. Second, the height that ramps lift the car is insufficient if you want to get right under the car – eg for exhaust work. Finally, to lift both ends of the car with ramps requires that you have four of them – and that you jack one end of the car up after you have driven the other end up one pair of the ramps. (And in this situation, ensure that the wheels on the ramp are on a flat section of the ramp and so cannot roll.) I have two sets of ramps and over the years I've used them literally hundreds of times – if you're on a budget and want to get the car up in the air just for engine servicing and the like, nothing beats them.

But what if you want to get under the full car length? In that case, if you still want to use ramps, you're into the area of full-length car ramps. Full-length car ramps can work in three ways.
1. Tilt ramps are of the sort you sometimes see at car

For low cars, normal ramps may be too steep in incline, causing the front spoiler to drag on the ramps. Making simple extensions like those pictured here fixes that problem.

When using jacks, jack-stands or ramps, always chock the wheels not lifted from the ground. You can use purpose-made chocks like this or simply offcuts of heavy timber. (Courtesy Toolstop)

dealers. The car is driven up the steeply inclined ramp and then, like a see-saw (teeter-totter), the ramps tilt forward to become horizontal. Props are then placed under the rear so that the ramps are stable. These are fine when someone else can guide you onto the ramp, but not so good when you are working alone. The 'dynamic' aspect is also harder to home engineer and build, and commercial ramps of this type can be expensive.

2. Inclined ramps that lead to a horizontal section the length of the car. These ramps don't require such a steep initial climb and, because they don't rotate when the car is on them, they are simpler to construct. But they are really long – typically twice the length of the car. I recommend this approach if you have plenty of space (eg you're working outside) or want to place a car up high for an extended period, where the labour of removing detachable entry ramps is minor in the overall scheme of things. These ramps are also expensive new, but secondhand units come up, often from wheel alignment workshops.

3. Inclined ramps the length of the car, that are jacked up at one end (usually the back) once the car is on them. These are my favourite full-length car ramps, because they can fit into a home workshop, can be made light enough that they can be handled by one person, and are safe. There is at least one commercial version of these ramps, and these ramps can also be made at home.

Note that both (2) and (3) above can be designed so that front wheel alignment turntables can be incorporated, increasing the versatility of the ramps. I use a set of ramps of the type described in (3), and I'll cover their home design and construction at the end of this chapter.

## CREEPER

If you are using ramps or jack-stands of any type you'll need a garage creeper – the wheeled platform on which you lie as you move around under the car. For what appears to be such a simple purchase, buying a good quality garage creeper is actually more complex that it first looks. First, the height should be as low as possible – that's especially important if you are using jack-stands or low ramps to support the car. Second, the design should use large, free-running wheels. Large wheels are needed because otherwise the creeper will jam when it meets anything on the workshop floor. The requirement for a low height, and yet large wheels, means that good creepers often have tricky designs where the top of the wheels are higher than the body of the creeper. The wheels also need to be castors, preferably rotating on ball bearings. Finally, it helps a lot if the creeper is a comfortable platform on which to lie.

Creepers should have large wheels and yet a low platform height. (Courtesy TraXion Engineered Products)

## HOISTS

There are good reasons why every commercial car workshop uses hoists. They are safe, fast to lift the car, easily adjustable for different vehicles, and can be set to any height. However, for home workshop use, they also have some disadvantages. They are expensive, take up a lot of space, and require specialised footings (and often a specialised power supply as well). You also need sufficient workshop height.

Hoists come in two fundamental types – two-post and four-post. A two-post design uses arms that extend outwards to the four jacking points located under the car. When lifted, the wheels of the car are therefore in free air, and so the suspension is at full droop. This allows brake and suspension work to be readily carried out. On the other hand, a four-post hoist is a drive-on design. The car is driven onto long ramps that are then evenly lifted into the air. The car therefore stays on its wheels. Four-post hoists are most often used in workshops specialising in exhaust work, as the clearance between the exhaust pipe and the

A two post, clear floor hoist. Specific footings need to be laid to install a hoist like this. (Courtesy Interequip)

(rebar), but often how this section is tied into the rest of the floor. The manufacturer will also have recommendations for bolting the hoist to the floor.

The next step, after ensuring you have adequate foundations, is to make sure that you can power the hoist. The available power supplies vary from country to country, so the advice here can be only general. The crucial point to note is that most hoists are designed for commercial workshop use, and so require workshop-level power – for example, three-phase 415V. If buying a hoist, especially a secondhand one, ensure that you can provide it with the power its needs. Note that while it is possible to change the electric motor for one that runs off a different power source, this can be a major, expensive exercise. Some hoists are designed and built specifically for home use – their primary advantage is that they will run off a household power supply.

I wrote above that a hoist takes up a lot of space. Given that the legs are actually fairly small in area, you may have wondered why I wrote that. The issue is that when you install a hoist, the flexibility of that area is much reduced. For example, it makes sense to install a hoist in a location that allows for the largest car you can ever picture using it. Therefore, the hoist is typically located in the middle of the bay, rather than towards one end of it. This, in effect, turns that whole bay into the 'hoist bay.' You cannot then use this area to (for example) lay out some steel fabrication on the floor, and, unless it's a long space, there won't be room to place storage cupboards at one end of the bay. If you have a very large home workshop, and there's not a problem in devoting one bay to just the hoist, then these problems

suspension is the same as when the car is on the road. A four-post hoist can still be used when doing brake and suspension work – for example, a jacking bar across the hoist can be used in conjunction with a small jack to lift the wheels.

All hoists place large loads on their footings – and this is especially the case for two-post hoists. A two-post hoist places major bending loads on the footings, and so requires that a deeper, better-reinforced section of concrete be used for this section of the floor. It is best if this is incorporated when the floor is being poured, but it is possible to cut out a section of floor and install a stronger section of concrete to carry a hoist. A four-post hoist spreads the load over twice the footing area, and is not as potentially unbalanced as a two-post hoist. However, with the car high in the air and (say) a long lever being used to undo a tight suspension bolt, there will still be plenty of bending loads being applied at the foundations of a four-post hoist.

If you are intending to fit a hoist of any type, carefully investigate the manufacturer's requirements for foundations. These recommendations will include not only concrete thickness, area and steel reinforcement

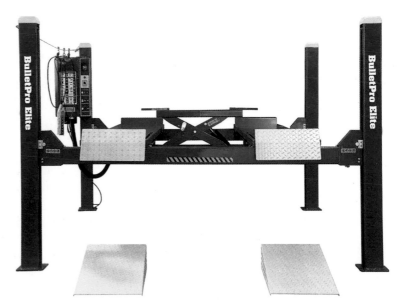

A four-post hoist that incorporates a lateral jacking platform that can lift one end of the car for suspension, tyre or brake work. (Courtesy Interequip)

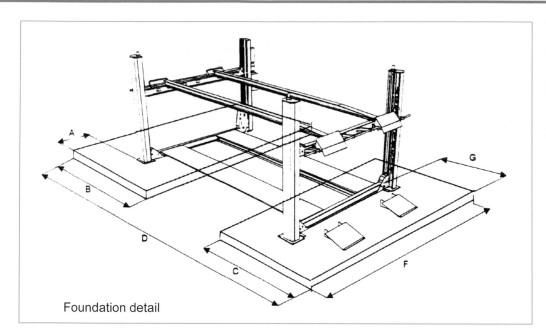

The specified foundation plan for a heavy-duty four-point hoist. The accompanying notes read in part: 'the area shown by the heavy outline must have a minimum concrete strength of $35N/mm^2$ and not be less than 150mm thick.' (Courtesy BWS Garage Equipment Ltd)

Foundation detail

won't occur. However, if you are tight for space, the hoist may intrude into the utility of the workshop more than you first expect.

There is a variety of hoist types available within the two- and four-post categories. There are those that use cables in conjunction with hydraulic cylinders, and others that use threaded rods on which nuts ride. Each type has its proponents, but there don't appear to be clear-cut advantages of one type over another. In two-post designs, there are hoists that have a connecting floor base between the two posts, and others that are free-floor but have a connection at the top of the posts. Free-floor is more expensive but getting the car onto the hoist is easier – there's no hump to cross with one set of wheels.

Despite my having a specific power outlet installed for a hoist, and having a strengthened section of concrete incorporated into my workshop floor when the concrete was poured, I've chosen to not use a hoist in my home workshop – I think it would tie up too much precious space. However, if I were to plan a new home workshop, I think I'd go an extra bay wide and plan to have that bay specifically for the hoist.

## PITS
In the days past, most home workshops had a pit – a sunken, narrow area under the centreline of the car, usually accessed by steps, or, in some cases, by a ladder. When not in use, the opening was covered in timber planks. Pits can be half-depth (you need to sit or squat to use them) or full-depth (you can stand while using them). I once removed and then refitted an automatic transmission in a car, working in only a half-depth pit. Pits have both advantages and disadvantages.

Compared with a hoist, a home workshop equipped with a pit does not need a high roof. Especially where local planning regulations prohibit high roofs, this is a significant advantage. Pits also have no running costs, whereas a hoist needs regular checks and maintenance.

On the downside, a pit cannot be moved to a new location, as can hoists or ramps. However, a more serious downside is that fuel vapour and exhaust fumes tend to settle in low areas – including pits. This can result in asphyxiation or explosion. To avoid these possibilities, intrinsically safe lighting systems must be used, and the pit needs to be force-air ventilated, for example through extraction vents located in the pit walls. Welding under the car while standing in a pit should never be attempted unless the pit has been completely purged of oil, flammable vapours and gases. Another danger is very simple – people falling into open pits. Finally on the list of disadvantages, a pit is not adjustable for width. This means that if you set the width to best work for a large car, a small car may have a track too narrow to allow the pit to be used. Conversely, if you set the width for a small car, a large car will have some underbody areas that are harder to access from within the pit.

It is expensive (but not impossible) to install a pit in an existing workshop, but it is easier and cheaper to install a pit as the workshop is being built. Pits can be constructed from concrete block walls built into place by a bricklayer, or from pre-fabricated welded steel sheet.

## ENGINE CRANES
Even if you only rarely need to remove and replace engines, an engine crane is a very useful tool to have in a home workshop. Most engine cranes use a hydraulic piston (like

A large commercial inspection pit. Note the intrinsically safe lighting, checker plate floor and safety markings. In a home workshop, add forced air ventilation as well. (Courtesy Premier Pits)

a large hydraulic jack) to lift a jib that can be extended as required. Long, splayed legs hold the crane steady. A typical low-cost hydraulic engine crane has jib extension positions for 1½ tons down to ½ ton, with the lowest load rating occurring when the jib is extended furthest.

When selecting a crane, look for one that can reach a good height (eg 2.2 metres, or a little over 7ft) at full height, maximum jib extension. Because the crane will spend most of its time in storage or tucked out of the way, you also want

A general purpose, low-cost engine crane. Useful for much more than just lifting engines in and out!

Using my engine crane to lift my mill for movement to a new workshop.

one that can fold down to compact dimensions. Even a crane capable of only half ton load at maximum extension will be adequate for nearly all home workshop use. As with jacks and jack-stands, when buying a crane, ensure it has appropriate standards ratings.

Over the years I've used my engine crane for lifting in and out engines, but also for moving my mill and lathe on and off vehicles and around the workshop. I even temporarily equipped the crane with a long jib and rear balance weights, and lifted into place the main roof beams of a home workshop I built.

## SPIT

A spit, sometimes called a rotisserie, is a metal frame designed to support your car's body from each end, holding it up in the air. Most designs comprise two large end assemblies joined by a long stabilising bar that passes under the car just above the floor. Spits are used on stripped car bodies to allow bodywork to be easily carried out on all fixed panels, including the floor and roof.

Spits can be home-made or purchased new. Occasionally, secondhand home-made items also appear. All spits have a means of supporting the car and a pivot axis around which the car can be turned. The car can normally be locked at any point in rotation. The most elaborate spit design incorporates two hydraulic jacks, one at each end. In addition, threaded rods are provided that allow the car attachment arms to be wound up or down. (Once the car's weight is correctly on the spit, these threads are locked and the hydraulic jacks do the lifting work.)

A different approach to a home workshop hoist. This hydraulic lifting platform has been integrated so that in its lowest position it is flush with the floor. (Courtesy Jack Olsen)

The reason for this apparent complexity needs a bit of explanation. The arms of the spit attach to the car, normally at the points where the bumpers bolt on. However, because bumper height varies from car to car (and sometimes even from the front to the back of the one car), the threaded rods are used to adjust the height of the attachment arms to match the car. These threaded rods have another, very significant function. If you imagine the axis on which the car is to rotate, and place an equal amount of the car's mass above and below that axis, the car will be able to be spun by hand. If there is more mass above the axis than below, the car will want to topple over – to always turn upside down. The threaded rods, together with an external jack (eg a trolley jack) can be used to position the car closer to, or further from, the pivot axis. This allows you to match the pivot axis to the distribution of mass in the car.

With this type of spit, once the car is on it, the screw threads must be adjusted so that the distance between the car attachment points and the pivot axis is the same at each end. (Again, when making this adjustment, you'll need to use a separate jack to take the weight off the threads, or alternatively, jack the car with the inbuilt jacks and then lower it onto some supports.) If you don't perform this vital step, which is not mentioned in some instructions, the car will be hard to turn, and may even break its mountings or damage the spit itself. (On spits without screw thread adjustment, the car must be manually jacked to exactly the correct height for attachment.)

With the car attached to the spit and set up correctly, the hydraulic jacks can then be used to raise the car to a height sufficient to allow it to be rotated without hitting anything on the ground. Each jack must be operated simultaneously – again, you don't want to bend your brackets or the spit.

If you buy a spit without hydraulic lifts or screw thread

A spit allows you to rotate a stripped body to make bodywork repairs.

adjustment (some homemade designs), it's likely that:

1. The car will need to be manually lifted a long way (eg 1.2 metres (47in) or more) before it can be attached to the spit at a position appropriate to the rotational axis and sufficiently high to let it fully rotate.
2. You will need to climb a small step ladder to do some of the work on the car (eg to work on the upper panels).

Before being placed on the spit, the car should be made as light as possible (no engine or gearbox, and preferably no interior, no doors, no glass, no suspension or wheels).

There are two areas of major weakness in a spit. The first is that the forces acting on the pivot points are very large, because, typically, the car is positioned 400mm (16in) or more from each pivot (any closer and you can't work on that end of the car). Any bending in the mounting brackets puts great leverage on the pivots and their supports. Secondly, the strongest attachment points to the car are normally the original bumper attachment holes, but getting sufficiently strong brackets to these holes can be problematic – there is often not much space. Remember that you are lifting the car from these points, and that the brackets need to be as laterally strong as they are vertically, to cater for turning the car body on its side.

In all cases, you will need to make your own high-strength, custom brackets – and so have welding facilities and the tools to cut heavy steel. Make sure when mounting these brackets that you use high tensile nuts and bolts with large diameter washers or load-spreading plates.

If you are working on a car body, it makes it so much easier to be able to set your own convenient working height and direction, and rotate the car to match that. For example, I cannot gas fusion weld upside-down, and I especially cannot fusion weld upside-down when working with very thin panels! But with the spit, you don't have to. The welds can always be beneath your hand. I have also found when panel beating, when making measurements (eg to assess engine swap possibilities), when removing fuel lines and altering brake lines, that a spit is invaluable. Finally, the spit I purchased is on large castors and so the car body can also be easily moved around the workshop.

## CONSTRUCTING CUSTOM FULL-LENGTH RAMPS

After using short ramps, a trolley jack and jack-stands for years, I decided that I wanted to gain better access to the

If you have a welder it's not hard to make these ramps. This picture shows the ramps in the inclined position. Note the intermediate supports placed under the inclined ramps, and the flat-topped stands on which the front wheels are resting. Eight pins lock together the stands and ramps. The jacking beam and jack stands can be seen, waiting to be used, on the ground to the right.

complete underside of the car. I'd fabricated and fitted an exhaust with the car elevated on only jack-stands, and found it very awkward. So when I started thinking about fitting a full aerodynamic undertray, I decided I wanted more working height. In addition, I wanted the ramps to be useable for wheel alignments – meaning the car needed to be level, with the front wheels on flat platforms large enough to take wheel alignment turntables. For storage, the ramps had to easily disassemble into pieces that were light enough to be handled by one person, they had to fit into my working space, and it would be helpful if they had two modes – a quick inclined mode for work like oil changes, and a horizontal mode for wheel aligning and full under-car work.

Some of the inspiration for the design came from the MR1 ramps developed in the UK. The MR1 ramps are described by the inventor like this. "When lowered, the system comprises two long shallow ramps supported by pivots at the upper ends. The lower ends of the ramps are connected by a rigid lifting beam. When a car has ascended the ramps, the lifting beam, ramps and car are raised together using a conventional trolley/floor jack, which slots into a central gap in the beam. The raised ends of the ramps are then supported by axle stands or blocks, and the trolley jack removed."

The MR1 ramps appear to be constructed from steel plate and angle, and the stands on which the front of the ramps pivot are made from steel angle. The MR1 ramps, however, do not have flat platforms at the front large enough to take wheel alignment turntables, and the ramps look quite heavy.

The ramps in the 'up' position. Note how this 2-ton vehicle is supported by the front wheels resting on the strong front stands, and the rear wheels by the four jack-stands (two each ramp). The ramps themselves are taking almost no bending load. To achieve a horizontal position, a jacking beam and a normal hydraulic jack were used to lift the rear of the ramps. Wheel alignment turntables can be used on the front stands.

After mulling over the issues for a while, I decided to take the following approach.

- Very strong and stable front stands, comprising flat-top, rectangular platforms supported on all four corners.
- Lightweight full-length ramps comprising ladder-like assemblies made from square and rectangular steel tube, covered on their upper surface with marine plywood.
- A jacking crossmember made from rectangular steel tube and heavily braced.
- Vertical supports (two at different heights for each ramp) that fit under and support the ramps, but only when the car is being driven up and down the ramps.
- A 4-pin connection system that locks the front stands to the ramps, initially with the ramps in the inclined position and then subsequently with the car in the horizontal position.

The jacking beam and ramps, shown before paint and the ramps' plywood top surfaces were applied. Note the strong gusseting used both top and bottom of the jacking beam.

The key point to keep in mind is that the inclined ramps need to be strong enough to support the weight of the car *only when the car is being driven on and off the ramps*. Once the car is on the ramps, the front wheels are on the wheel alignment stands, and the back wheels are positioned part-way along the ramps. If the jacking beam is placed under the rear wheels (or near to that point) the ramps no longer have to take the weight of the car. This means the ramps can be made much lighter, allowing them to be handled by a single person. Don't forget that to give the ramps the required stiffness when in the inclined mode, supports are always placed underneath – as can be seen in the photo on the previous page.

An important safety aspect of the ramps is the locking pins. Initially I was going to simply use pivots to connect the inclined ramps to the front stands. However, if you were then reversing the car down the inclined ramps and you braked, the rear of the ramps could slide backwards and the front

The front stands. Note the basic cube-shaped framework that is strengthened by diagonals on the side and front faces, and two smaller braces on the front face.

stands could overturn rearwards. Without the pins, if you had overshot the front stands and hit the buffers, the front stands could have overturned forwards. Using four pins – and not just two pivots – keeps the structure locked together.

The pins work like this. When the ramps are in the inclined position, four pins are inserted on each of the front stands. During jacking of the rear of the ramps, the two upper pins are removed to allow the ramps to pivot on the lower pins. When the ramps are horizontal, the upper pins are inserted through new holes that again lock the assembly together.

The finished ramps are strong enough to support a

2-ton vehicle, yet can be easily disassembled and then stored by one person working alone. (The heaviest parts are the ramps, that can – just – be handled by one person, although using two people makes it easier.) As mentioned, the front stands can hold the wheel alignment turntables, or when these are not needed, wooden spacers can replace the turntables if it's desired that the car stay perfectly level. While more expensive to build than normal short ramps, the cost is also vastly less than buying a hoist – and the workshop space remains completely flexible. However, I do need to add that building the ramps is quite a lot of work.

### Safety!

Before even thinking of constructing ramps like these, you must be an experienced welder, and have a good quality machine that can make consistent, strong welds. This is not a project on which to learn to weld! Another safety warning – the ramps must be used as designed, initially with the underneath supports in place, and the four pins per ramp inserted to hold together the structure in the inclined position. When the car's front wheels are on the wheel alignment stands, the upper pins should be removed and the jacking beam placed under the rear wheels. Jacking then occurs to a horizontal position, jack-stands are placed near to the jacking beam, the ramps lowered onto the jack-stands, and the upper locking pins inserted in their new holes. *If you find any of this confusing, or feel you may get the sequence wrong, do not build these ramps!*

The ramps in the inclined position, with the upper pins removed to allow the angle between the ramps and stands to change as the rear of the ramps is jacked-up. Note that the bottom pins always stay in place – these are the pins on which the ramps pivot. If you want to make it easier to remember this, you could paint these lower pins red – when there is a car on the ramps, these are the pins you never remove!

Four pins inserted, locking together the system in the 'up' position. Remember that there are always four pins in place on each of the front stands, except when jacking the rear of the ramps.

### Materials

I am not presenting a formal plan or step-by-step instructions for making these ramps. If you have the capability to build them, then you'll be able to work it out quickly from the pictures and the following dimensions. Because aspects such as tube wall thickness are very important to strength, I am not providing the tube measurements in 'approximate Imperial,' as I have done for other conversions in this book. If you are working in a country with Imperial measurements, ensure that the Imperial tube size equivalent is not smaller in either overall dimensions or wall thickness.

The front stands are 480mm (19in) high, 490mm (19¼in) wide and 405mm (16in) deep, plus another 170mm (6¾in) for the front buffers that extend forwards of the vertical supports. The buffers are 100mm (4in) high.

The stands are made from square tube of 40mm x 2mm wall thickness. The supporting props (used when the ramps are inclined) are made from the same material.

The ramps are 3.7 metres (12ft) long and 400mm (15¾in) wide. They are made from rectangular tubing placing on edge, 75mm x 50mm x 2mm wall thickness. The cross pieces are of the same material used for the front stands, and are spaced at 300mm (12in) centres. The top surface of the ramps is 12mm (½in) marine plywood – do not use lower grade plywoods as they are not strong enough.

The jacking beam is made from rectangular tube placed on the flat, 100mm x 50mm x 3mm wall. It is braced on top by 40mm square tube (2mm wall) and on the underside by offcuts of the 100 x 50 x 3mm tube. The jacking beam is 2050mm (81in) wide – make the centre height and width to suit your hydraulic jack. The pins are 12mm in diameter, and have 16mm handles welded to them.

**Chapter 7**

# Approaches to innovative design

To make best use of your home car workshop, not only will be you be performing car service and maintenance jobs, but it's likely that you'll also be coming up with your own unique approaches to making things. Those 'things' might be car modifications, or they might be items that you're making for use in your own workshop. In this chapter, I want to look at what I have found to be the best approach to design and creation. This is important, because I think many people are stifled in their home workshop creativity – in part, because they listen too much to others!

## BACKGROUND

For about 15 years or so, my full-time job was writing about technical – and often engineering – matters. Very often, that meant designing and building the unique projects that I was writing about. Sometimes those projects were mechanical, some were electronic, and some were pneumatic. Most were also automotive in nature – air suspension with a custom electronic controller, a new design of turbo boost control, tuned mass dampers to reduce light vibration when driving, and so on. So how was I to know the best way of developing such projects?

In my home library I have lots of books on engineering, and I've read and understood at least the broad thrust of all of them. But it's always seemed to me that engineering books ignore what is surely the most important aspect of

engineering real-world solutions to problems – and that's devising new and unique solutions. The books are full of theory, full of solved problems, full of examples where the solutions are just a few equations away.

But unfortunately, in the real world – and especially for people being creative in home workshops – these approaches are often a dead end. That is not to say that the books are of no use – emphatically, far from it. But it is to say that, if you're confronted with an engineering problem, the solutions may initially come from your head – and not books. Instead, let's take a look at what approaches can provide immediate and sound benefit when you need to devise solutions to engineering problems, and then put it into practice with a real-world example.

### Step 1 – Not being constrained in thought

One of the impediments in devising solutions to engineering problems is that thought processes are often constrained along narrow lines. Engineering textbooks are terribly bad in this regard – for example, car suspension textbooks are just filled with the suspension designs that are already widely used in cars. A car suspension textbook won't mention railways or horse drawn wagons or bicycles. It won't mention aircraft undercarriages, or even sports shoes. But if you want to devise unique solutions, you need to think of the topic in the broadest possible way.

Genuinely innovative engineering design like Iron Bridge in Shropshire, England, the world's first bridge constructed of iron, did not come about by following what others had done! An important part of innovative design is to listen to what others say, but to then sometimes ignore them. Viewing a structure such as this really makes you reconsider what constitutes a ground-breaking design, and the thought processes that led to it.

Suspension, for example, is fundamentally about springing, damping and linkages. In more complex configurations, it's also about interconnectivity – for example, of wheels. If you examine the topic as widely as you can, you'll gain a fair better understanding of what is actually possible. I am talking here about innovative design – not just changing springs in an otherwise stock car. If you're trying to devise a suspension that, for example, is ultra-lightweight, has long travel, and uses hydraulic damping, you should first think laterally about *all the devices that you can think of that use suspension*.

For example, as well as cars, also think about:

- Railways – linkages, interconnectivity, lateral springing, steel springs, rubber springs, air springs, hydraulic damping.
- Horse-drawn wagons – large diameter wheels, lateral and longitudinal springing, timber springs, steel leaf springs, friction damping, motion ratios.
- Bicycles – motion ratios, air springs, steel springs, plastic (elastomer) springs, hydraulic damping, friction damping, air damping, oil damping, linkages.
- Aircraft undercarriages – interconnectivity, rubber springs, elastic cord, nitrogen springs, hydraulic damping, linkages.
- Sports shoes – hysteretic (internal deflection) damping, plastic springs, air springs.

And it's not just suspension. If you're trying to devise a new intake manifold for an engine to take advantage of tuned pulse behaviour, you need to look not just at car (and other) engines, but to also consider the way that sound waves behave. Here's where a textbook is great – not one on engine intakes, but instead one on sound waves, eg how they travel and reflect. Most 'old-fashioned' textbooks on sound will begin with the fundamental ideas – ideas that you can constantly be thinking about in the context of engine intakes, not the musical instruments often used as examples.

Limiting yourself to pursuing knowledge only within the narrowly defined area of interest means that you immediately constrain yourself to taking the path that others have already done: in other words, innovation will be lacking.

### Step 2 – Identify and prioritise the elements of the problem

Any engineering solution for a problem will achieve some outcomes better than others – you can't solve everything at once to the same degree. Therefore, break the design down and then prioritise which problems are most important. For people working in a home workshop, often 'cost' and 'ability to build the design' are very important limiting factors. If you can't afford it, or can't make it, then the solution is clearly a failure for you. But after these factors, what are the next most important?

Prototype Leatherman tool, that I photographed at the Leatherman factory in Portland, Oregon, USA. A good example of a design made by someone who deliberately chose to go their own way.

At the time of writing, my current project is the front suspension design on an ultra-lightweight (20kg/44lb) vehicle. The priorities are:

1. Able to be built, and affordable.
2. Low weight.
3. Stiff in roll.
4. Suspension travel of 100-150mm (4-6in).
5. Sufficiently durable to cope with 40kg (around 90lb) wheel loads and 1g bump acceleration.
6. Able to be used with steering that can achieve no bump steer, and have Ackermann geometry.

It was only when I realised that 'stiff in roll' was so high up the list that I started to radically change my design ideas.

If you're engineering a bracket, what's more important? Weight? Strength? Ease of manufacture? Appearance? Corrosion-proofness? Or does the bracket simply have to be made with materials you have at hand, right now?

Breaking the problem down to its component parts and then prioritising criteria lets you see the trees in the forest.

### Step 3 – Consider the views of others ... but often, ignore them

Especially with the speed, reach and access of the Web, these days you can always find someone to discuss ideas with – no matter how esoteric the technical field. You might want to modify your Prius, fit Weber carbs to your big block, or restore your Tropfenwagen (I wish!). Whatever the automotive project, you're sure to be able to find someone online to talk to. That can be of great value – but more often than not, unless you apply a good filter, *it can stop you succeeding.*

Consider, for a moment, a group of people. In that group there'll be only a few who'll be really smart. That's a politically incorrect thing to say, but it's true – bell-shaped curve and all that. Therefore, the feedback you get on any idea from such a group is quite likely to be not well thought out. It may even be misleading, wrong, stupid, slightly incorrect, ridiculous – and so on. In fact, if the group with which you are communicating is typical, and your idea is new, *the feedback is much more likely to be wrong than right.*

And it doesn't matter how many in the group say it: after all, by definition, if you're coming up with a new idea or a new approach, they haven't already thought of it. And, human nature being what it is, people tend to reject new ideas.

Over the years I have received some excellent advice from online people I've never met. But you really have to cull the vast majority who are simply not equipped to give you useful advice. The 'old fashioned' ideas like assessing the background of the person, finding out what they have achieved in the field you're discussing, considering how lateral their thought processes are, and

ascertaining whether they've ever actually carried out the idea (or a variation of it) – all must be employed if you're to get advice that is of use. That's a very simple notion – don't listen to idiots – but I see so many people who come up with a great idea, then retreat, tail between their legs, when other people dump on it from a great height. Remember also, that the more expert someone is at a topic, the more they tend to accept the prevailing wisdom. Therefore, ideas that challenge that wisdom are more likely to be rejected.

Let's look at an example – you've decided that insulating the full length of the exhaust pipe in your car will keep the gases hot and so make them flow faster, improving engine power. You ask for some feedback from a Web discussion group and people come back with:

"Won't work, don't waste your time." (This person has contributed nothing; ignore them. Totally ignore them, don't even let their comments penetrate your consciousness.)

"That approach was tried in the Sixties, it didn't work." (Nearly as pointless – no evidence, no detail. Maybe ask them for some detail – invariably, you won't get it.)

"I tried wrapping my extractors [headers] and I got only a 2 per cent gain on the dyno – so I decided it wasn't worth it." (Very useful – ask more questions.)

"I did this once and the drop in engine bay temp was really good. Before I did the insulating, the engine bay was stinking hot; afterwards, it was clearly a lot cooler. Fuel economy seemed to go backwards though." (Very useful – ask more questions.)

"The heating of the exhaust gases will cause greater

One of Charles Parsons' model hull designs, made while he was developing the hull shape of the fastest boat in the world in the 1890s. All the test models were powered by rubber bands – an excellent, effective and low-cost approach. I photographed this model in the Discovery Museum, Newcastle upon Tyne, in the UK. The full-size ship is nearby...

refractive index, which in turn will cause the speed of sound to alter. This causes rarefactions of the exhaust gas that will lead to poorer flow. Or that's what my old fluid dynamics lecturer told me, if I remember right." (Absolutely typical of the rubbish I see all the time. Totally ignore them.)

"If that would work, why doesn't everyone do it?" (This sort of response enrages me: often the answer is that no-one's ever thought of doing it …)

Remember, if you're trying to devise unique solutions to problems, by definition no one else has done the same before (or anyway, the solutions aren't widely known). Therefore, don't expect support for your ideas – you usually won't get it.

## Step 4 – Test

A simple test is worth any textbook full of theory. Engineering theories are simplified models of the world; in many cases, so simplified that they are relevant only when other factors are constrained to a silly degree.

Engineering mathematical analysis can show that a bridge won't fall down. But no it can't: the maths will tell you nothing about the homogeneity of the steel, the quality of the labourers, the presence of terrorists who next year destroy that bridge. Of course the maths can't tell you that, you say. So, in fact, the maths cannot tell you whether the bridge will survive or not – it can tell you only about *one aspect* of its structural design. The example may be a little silly (engineers don't claim that the maths will tell you about subsequent terrorist attacks) but the principle remains sound: engineering theory is heavily constrained by the elements used in the model – include all elements that actually apply in the real world, and it very often gets too hard.

That's why testing is so good – it shows you *what is actually occurring*, not what is predicted to occur. If you are exploring a new solution to an engineering problem, do tests as early in the process as you can. Try also to do the tests in the simplest, cheapest way. Simple, cheap tests do not have to be poorly thought out or ineffectual. They may give uncertain answers (fine, test more thoroughly), they may give a really positive result (great – keep developing!) or they may give a really negative result (great – don't waste any more time!).

People often seem terribly reluctant to do any testing, preferring instead to theorise. But because the test determines what actually happens, and theories are just simplified models of reality, if it's at all possible, test, test, test. So if you want to know what insulating the exhaust can do, it's worth spending the money and wrapping the thing – not spending days or weeks in Web discussions, reading thermodynamic textbooks, talking to learned professors, etc, etc. *Just do it and find out!*

In the case of my ultra-lightweight vehicle front suspension, after thinking about it for a while, I built the design. The testing – subjecting it to specific bumps, cornering on a skidpan against a stopwatch, data-logging vertical accelerations, taking action pics of full-load cornering – showed many things were occurring that I had never even considered. Some of those things were good, some of those things were bad – but now I know exactly what actually happens in the real world.

In some cases, of course, testing cannot occur of the full design – it's too big or too expensive a project to build just to test. In that case, see if you can model the results. After consulting theory textbooks on the geometry of the required Watts link for my vehicle suspension, I built a wooden model and tested it to ensure the geometry was correct.

In electronics and pneumatics, it's often possible to rig up a 'quick and dirty' project, and then test it to see if the results are as expected. I once spent years working with an electronics engineer who had designed literally hundreds of projects, many of them unique. His recurring refrain was: "Nearly always, something unexpected happens when you test the prototype." Never mentally commit to an approach until there's been a demonstration – of some sort – that it will work.

Here's an example. Using plastic sign material and gaffer tape, I trialled an under-tray on the front of a Toyota Prius. It worked well – measurably improving fuel economy through a decrease in aerodynamic drag. Based on the positive results of that test I then made the 'proper' version from engineering plastic. On the other hand, I installed another trial under-tray under the back of a Honda Insight, but this time the tests showed that it made no difference – so I took it off and the project proceeded no further. After the Prius results, I'd been expecting the Honda under-tray to work, but it didn't and so I was glad that I hadn't spent the time and money building a 'proper' under-tray.

## AN EXAMPLE: BUILDING A CUSTOM AIRBOX

The above section describes an approach to innovative home workshop design, and I'd like now to use an example of this process. The project was the design and construction of a new airbox, needed as part of the turbo conversion of my Honda Insight.

## Step 1 – Not being constrained in thought

The first part of the approach is to be not constrained in your thought processes – so in the case of the Honda's new airbox, what other vehicles are fitted with airboxes?

I could have looked at airboxes fitted to trains or aircraft, but in this example, road vehicle airboxes are already very well optimised for their application. However, in addition to

This custom airbox was developed using the design process described. It needed to filter well, fit in the available space, have inlets and outlets positioned appropriately, flow well, use a commonly available filter, and be straightforward to make in my home workshop.

looking at the designs of airboxes on lots of different cars, I also looked carefully at motorcycle airboxes (the Honda has a 1-litre engine). But I found that motorcycle airboxes were the wrong shape, and were also smaller than I preferred.

And other cars? Despite looking at literally dozens of different compact cars, I couldn't find even one 'factory' airbox that fitted into the space, and had the inlet and outlet in the right locations. To get what I wanted, I needed to make my own airbox.

### Step 2 – Identify and prioritise the elements of the problem

When making my own airbox, what were the criteria it needed to match? In order, I decide that the airbox needed to:
1. Fit in the available space.
2. Filter well.
3. Have inlets and outlets positioned appropriately for the application.
4. Flow well.
5. Use a commonly available filter (why make life hard and expensive for yourself?).
6. Be straightforward to make.

So how to fit a good quality (and that means a standard paper element – more on this in a moment) filter with a large area into a small volume? One air filter shape that would fit in the available space was a long and thin rectangle. But this would have needed in turn a long, thin box that separated into two halves (so that you could change the filter) – and such a design would be pretty difficult to construct. And anyway, a long and thin filter gave a small filter area.

So after lots of thinking about different options, the following design was developed:
• large cylindrical paper filter element, 290mm (11½in)

long x 118mm (a little over 4½in) in diameter, normally fitted as standard to a Saab 9000
• cylindrical airbox, made from 150mm (6in) diameter, thin-wall truck steel exhaust pipe
• flat end plates, cut from steel sheet

This approach looked like it would match all the above criteria.

### Step 3 – Consider the views of others ... but often, ignore them

In the case of the new air filter, I could have looked at what other people have done and followed them. In fact, there

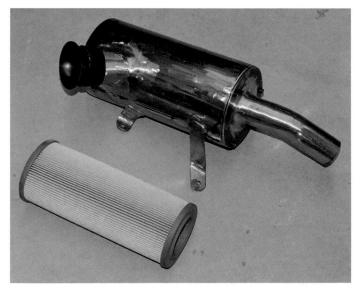

The airbox was constructed of 150mm (6in) thin wall truck exhaust tubing. The huge filter that it takes is normally fitted to a Saab 9000.

are just a few other people who have turbocharged a Gen I Honda Insight. They have either retained the standard airbox (with convoluted plumbing required from it to the turbo) or have replaced the factory air filter box with a cone-type aftermarket filter.

Testing over the years has shown that the best filtration invariably occurs with a factory-style paper element, and so I wasn't much interested in aftermarket oiled cotton filters and the like. Keeping the standard filter box was an option, but its inlet and outlet were positioned in the wrong locations to match the turbo application. Those who used cone-type filters had positioned them in the engine bay, breathing hot air. So there was little point in asking what others had done – I could see what they had done, and in every case it was a poor job compared with what I wanted to achieve.

### Building the airbox

The design was MIG welded together (with just a little bit of brazing as well). The starting point was a 300mm (12in) length of 150mm (6in) diameter, thin-wall truck exhaust pipe. The 150mm (6in) tube gives about 15mm (just over ½in) of air space all around the cylindrical filter.

At the outlet end, a ring was cut from thin steel plate and then welded into place. This ring was then drilled and nutserts installed. (The nuts could also have been brazed into place prior to the ring being installed.) A 60mm hole was cut in the removable lid of the airbox. To aid airflow from the filter into the exit tube, a plastic bellmouth was glued into place. The bellmouth was cut down from a flared port salvaged from an old subwoofer enclosure. In addition to the glue, the bellmouth was held in place by being firmly sandwiched between the filter and the lid.

A cone adaptor was formed in steel exhaust tube to adapt the larger airbox outlet to the 50mm (2in) tube that, in this case, runs to the turbo inlet duct. The adapter was formed using a cone-shaped steel mandrel (as it happens, a plumb-bob!) and my hydraulic press. In this installation, an unusually long exit tube was welded to the lid. The airbox lid was held in place with three screws that passed into the nutserts installed in the inner ring. A thin rubber gasket sat between the lid and the ring. At the other end of the airbox, a steel endplate was welded into place. To locate the filter at this end, three small steel spigots were brazed into place prior to the endplate being attached. The filter will insert fully into the airbox only when the far end is nestling within these locating spigots.

The air intake comprises a 63mm (2½in) tube cut at an angle and then welded to the wall of the airbox tube. Prior to welding, an oval-shaped opening to suit the tube was cut in the airbox wall. A plastic flared extension is connected to the intake via a short length of rubber hose. This flare is actually the internal end of the subwoofer vent tube that also provided the bellmouth airbox exit. Three mounting brackets

**The airbox being trialled in the car for fitment. Intake air is picked up from behind a headlight. Measured flow restriction is close to zero.**

were welded directly to the airbox wall. The final product was blasted and then powder-coated.

### Step 4 – Test

In this case, testing could be conducted only after the project was complete. So how did the airbox perform? The answer is very well. The measured airflow restriction on the airbox outlet was just 3in of water – that's 0.1psi. Filtration? I haven't tested it but I am confident that a factory air filter will be working very well. In fact, the filter is so huge for the engine that I doubt it will ever need to be changed – so far, after about 10,000km (6000 miles), I can see no dirt build-up at all.

### Summary of approach

Lots of people follow what others have done, rather than thinking through the process for themselves. In the case of the new airbox, I achieved an outcome that works better than any other approach I've seen done on the same car (not that there have been that many of them!). Significantly, it's also not like any other home-made airbox I have ever seen.

I think you'll achieve best results in your home workshop projects – especially those that are unique – if you follow this approach:
- Step 1 – Not being constrained in thought
- Step 2 – Identify and prioritise the elements of the problem
- Step 3 – Consider the views of others… but often, ignore them
- Step 4 – Test

In short, if you've thought long and hard about an approach, back yourself …

**Chapter 8**

# Welding

Whenever you want to permanently join metals, welding is usually the best choice from the perspectives of strength, durability and appearance. Cars have been welded since all-steel monocoque bodies started dominating at around the time of World War II, and welding, in one or more of its different forms, is used in nearly every metal product now made. A factory producing metal goods that doesn't use at least some form of welding or brazing – whether that's arc welding, MIG welding, TIG welding, gas welding, resistance welding, spot welding, furnace brazing or one of the other types – is almost unknown.

In my workshop, I have a proliferation of welders. Firstly, I have an oxy-acetylene welding kit. Where I live, that's a rather expensive system to have. This is because gas cylinders are hired rather than bought outright, and so there is a monthly charge – even if you're not using any gas. However, the oxy is perfect for brazing and heating to bend or heat-treat metals. I find the oxy flame a delight to use, with its ability to perform very fine work (eg silver-soldering brass fittings together), joining thin-wall chrome-moly steel tube by nickel-bronze brazing, and fusion welding steel intercooler and exhaust plumbing. I also prefer to use the oxy on car bodywork repairs – it's slower than MIG welding but it produces welds that require much less grinding back, especially if you want the panel to look good on both sides.

I also have a 200-amp MIG welder. This is my favourite – a welder than can work on thin sheet right through to heavy plate, and one that's very easy to use. If I had to have only one welder, it would be the MIG. The MIG is my general-purpose welder. In just the last week I have built a set of full-length car ramps, welded together some steel brackets, made some steel shelving, and repaired a clamp.

I also have a spot welder that I use for welding thin sheet together. This is not a tool I use as often as I expected to, but its last use was in constructing a tray to hold a car battery that was being relocated to the rear of the vehicle.

And finally, (and although it's not a welder, I think it should be included here), I have a plasma cutter, a tool that's very useful. It's only a relatively small unit, but I use it to cut thin sheet. It's also excellent on bodywork repairs.

And my first welder? It was a secondhand arc welder – now sold. I use one or other of the welders – and it's usually the MIG – nearly every day I am in the workshop. I'll start here with the cheapest of welders – arc.

## ARC WELDING

Arc welding is called arc welding simply because it uses an arc (unlike MIG, 'arc' is not an acronym). Buying an arc welder is the cheapest way to get a welder. New arc welders are available at home budgets, while secondhand units – often complete with gloves, a helmet and slag

### NOMENCLATURE
In this book, I have decided to use the most common names for each of the welding types. However, each of these welding types is also known by other names. The table below gives some of the variations.

| Name used in this book | Alternative names for same process |
|---|---|
| Arc welding | Stick welding<br>SMAW (shielded metal arc welding) |
| Oxy-acetylene | Gas welding |
| MIG (metal inert gas) welding | GMAW (gas metal arc welding) |
| Spot welding | Resistance welding |

In addition, there are flux-cored arc welding (sometimes called gasless MIG), and gas tungsten arc welding (TIG) welding, that are not covered here.

chipping hammer – can be found very cheaply. For general purpose welding – making workbench frames, building shelving, making a heavy-duty underbonnet bracket or building a trailer – nothing beats an arc welder for price.

However, this type of welder is best suited for welding mild steel that's in the range of 2-8mm ($\frac{5}{64}$-$\frac{5}{16}$in) in thickness. (Yes, thicker steels can be welded, but normally it requires higher currents than are available in the cheaper welders.)

An arc welder is basically just a big AC transformer.

**Arc welding a turbo manifold together. It's made from thick wall ('steam pipe') fittings.**

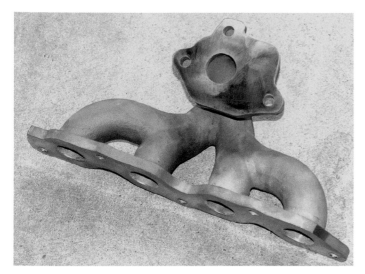

The completed four-cylinder turbo manifold. Grinding-back the arc welds has given the manifold the appearance of a casting.

A compact transformer-type arc welder. Welding current is adjustable from 40-100 amps. (Courtesy Machinemart)

A 75A inverter-type arc welder. (Courtesy Craftsman)

Like any transformer, two windings are used. Mains (line) voltage at 10 or 15 amps is applied to one set of windings, while the others generate a much lower voltage at a much higher current – say 120 or more amps. (Inverter welders use different technology to achieve the lower voltage/higher current.) The welding current is adjustable and is usually marked in 'amps.' In the simplest of welders there are no other controls except for an on/off switch, while others have a control that can vary current.

To operate an arc welder, one wire (the earth or ground wire) must be attached to the material to be welded, or to a metal table on which the material is placed. The other cable runs to the hand-piece, in which the welding electrode is clamped. When the electrode touches the work, electrical current flows from the welder through the rod and workpiece back to the welding machine. If the current flow is established, and then the electrode quickly pulled back a little from the workpiece, a very high temperature arc forms between the end of the rod and the workpiece. This melts both the welding rod (which becomes filler material) and the materials being welded.

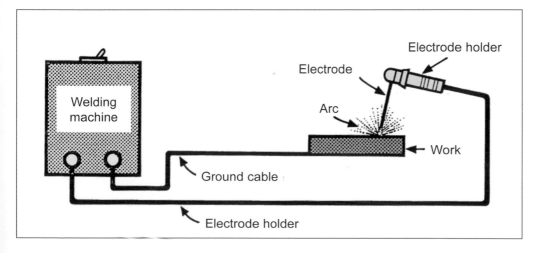

The basic circuit of an arc welder. The transformer (or inverter circuitry) produces a high current at a low voltage.

## Rods

The welding electrode (the stick or rod) has an inner core of metal similar to the material being welded. This core also has a diameter that is proportional to the material – as the workpiece gets thicker, so too should the rod. The inner of the rod is surrounded by a welding flux. When the molten material solidifies, the flux forms a separate layer on top that can later be knocked away with the chipping hammer. The flux on welding rods serves these functions:

* provides a gas shield
* gives a steady arc by providing a 'current bridge'
* cleans the surface and slows the cooling of the weld
* introduces appropriate alloys into the weld

In addition to the differences in diameter described above, welding rods vary in other characteristics. (Despite 'general purpose' welding rods being sold at every hardware store, in fact rods should always be matched to the application.) Specific electrodes are available for welding:

* mild steel
* cast iron
* stainless steel
* copper, bronze, brass, etc
* high tensile steels

Don't underestimate the importance of using the correct electrode. On a cast iron bracket I once had welded, even a welding rod designed for cast iron gave a result so poor that a single hammer blow broke off the (apparently sound) weld. On a high tensile steel tube, a tack made with welding rod designed for mild steel had less strength than a blob of glue.

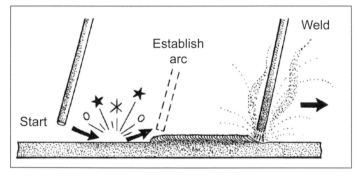

When beginning arc welding, the most difficult skill to gain is striking the arc. The end of the rod is scraped along the workpiece and then lifted a short distance. As the weld proceeds, the rod is consumed, so the welder's hand must gradually move closer to the workpiece.

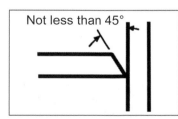

When working with thick material, a bevel should be first ground so that full weld penetration occurs.

## Joint preparation

Joint preparation is very important in gaining a high quality, strong weld (this applies to all welding and brazing techniques, not just arc welding). A joint that has an overly wide gap, or is dirty, rusting or greasy will not give a good result. The way in which the workpiece is designed will also influence the structural integrity of the finished item. So let's take a look at the different types of commonly welded joints.

A butt joint occurs where two pieces of metal are offered up to one another. They are on the same level and a small gap is left in between. The weld fills the gap, penetrating through the thickness of the sheet. Where access is more difficult and/or the material is thicker, the edge should be ground away to allow better penetration. This is called a single V butt joint. Material that is thicker again, and which can be accessed from both sides, can use a double V butt joint, where material is bevelled away on both sides.

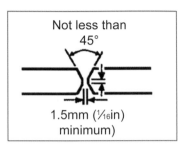

**Preparation of a welded butt joint in thick material. Note the ground bevels and the small gap left between the pieces.**

A lap joint, where the two pieces of material overlap each other, can be a much stronger join than a simple butt joint. Not only is the weld guaranteed full penetration on the exposed end of the material, but holes or slots can also be cut in one sheet to allow plug welds (through a drilled hole) or slot welds (through a cut slot) to be made. With appropriate penetration, these welds can be very strong.

Another type of join is a T-joint. As with butt joints, the material can be bevelled to allow better access and weld

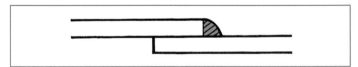

**A lap joint. When welding this type of joint, put more heat into the lower sheet to achieve best penetration.**

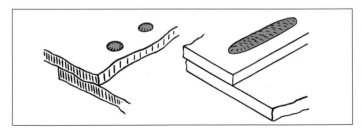

**Plug welds (left) and slot welds (right) are made through openings previously made in the upper sheets. If you do not have a spot welder, plug welds can be used as a good alternative.**

penetration. A fillet weld is used where surfaces join at 90 degrees. This type of weld can use multiple beads, the latter being especially effective if the surfaces are bevelled to allow better penetration. Finally in common welds, there are edge welds.

When designing a component that will be welded, the type of welded join that will be used should be carefully considered. For example, if the design is such that access can be achieved only to one side of the workpiece, some joins may be unnecessarily weak.

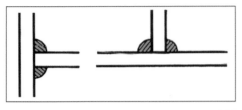

A T-junction weld needs more current than a butt joint. This type of weld is a good one to practice on scrap first.

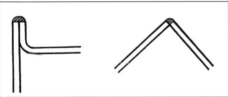

Edge welds require low current or it's easy to melt away the edge.

## Technique

One of the hardest aspects that a beginner finds in arc welding is striking the arc. The arc must be struck rather like a match is lit. Scraping the end of the electrode across the work, and then rapidly lifting it slightly, is how it's done – but it's often not as easy to do as it is to write! Practicing striking the arc on a piece of scrap, and adjusting the current and seeing the affect this has on arc striking, are the best two approaches. The arc should be the shortest that will produce a good weld (eg 1.5mm (¹⁄₁₆in)), so be careful not to lift it too far after the arc has started.

The textbooks define the rate of forward process and the angle that the rod should be held to the work, but much of the fine detail of this depends on the material being welded, the current and rod being used, and the skill level of the operator. One trick is to learn to look beyond the arc to the pool of molten metal itself – that tells you a lot more about what is going on than the appearance of the arc itself.

It's also important to know what good and bad welds look like, and the reasons why bad ones are like they are. The diagram on the right provides a good start in recognising errors – note that both plan and cross-sectional views are shown. Cutting a welded practice piece in half and assessing the penetration of the bead is worthwhile.

Another way to see what is happening is to deliberately weld in error. What does a weld look like when the current is much too high, or far too low? What is the bead's appearance when welding excruciatingly slowly, or far too

Excellent practice at holding the welding handpiece steady and at a constant distance from the workpiece can be achieved by using a sharp pencil, a washer and a piece of paper. Practice pushing the washer along without letting the washer slip out from under the pencil or the pencil touch the paper. When doing the exercise, position yourself in your normal welding position, eg sitting at a bench.

rapidly? If you deliberately create the error, you'll be able to recognise what's going on when the bad weld appears, even when you're trying hard to do it right!

Running multiple beads, one over the top of the other, and weaving from side to side are often suggested in welding textbooks. If you're a competent welder (and in the case of the multiple beads, make scrupulously sure that the slag is removed after running each bead), no problem exists in doing these things. But if you're a beginner, multiple beads and weaving with an arc welder is a recipe for slag inclusions and voids.

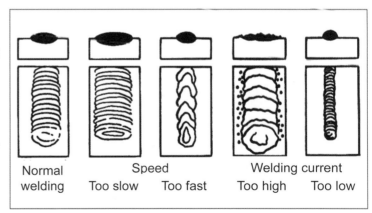

Arc welding diagnostics. Note that both the normal and cross-sectional views are shown. The best way of assessing penetration is to cut a sample weld in half.

## OXY-ACETYLENE WELDING

Go into any commercial workshop in the country, large or small, and you'll almost always find an oxy welding kit. There'll be the black bottle (oxygen) and the smaller maroon bottle (acetylene), some hoses and a handpiece. But isn't an oxy set a bit old-fashioned? After all, these days you can buy cheap MIGs, plasma cutters – even TIG welders. So what use would have an oxy welding kit have? The short

An oxy-acetylene welding kit. Add oxygen and acetylene cylinders, and you're ready to go. This set includes the regulators, handpiece and a cutting attachment.

answer is: a lot! An oxy-acetylene welding kit can do all the following:

- fusion weld
- braze weld
- silver solder
- heat metal to allow it to be bent and formed
- heat metal to allow it to be hardened or softened
- cut metal

That is an incredible list! Why? Well, before we get into the nitty gritty, here are some real world uses.

- Take the first – fusion welding. In brief, that's where you melt together metals of the same sort (eg steel), usually with the addition of small amounts of a filler rod made from the same metal. I fusion weld steel intercooler plumbing.

- Brazing? Again, it works well on intercooler plumbing, and I recently cut and brazed an oil return line from a turbo. Brazing doesn't require that the metal is melted – just heated to a dull red. As a result, there's less distortion and the process can be used on very thin metal. Brazing can also be used to join dissimilar metals – eg copper to steel.

- Silver soldering? I've used this on a turbo conversion – in that case, on a high-pressure oil supply fitting. Silver soldering (nothing like normal soft soldering) is like brazing except it uses a rod containing silver, and is good for very close-fitting parts.

- Whenever metals are bent, for example when making a bracket, hot bending subjects the material to lower stress than cold bending. The thick steel bar that needs a huge hammering in the vice when it's cold becomes the bar easily bent with some gentle taps when it's hot.

- By heating steel to different temperatures (usually indicated by the colour of the material) and then quenching it in different baths (eg oil or water), steel can be hardened. Also, materials that work-harden (eg copper) can be softened by heating and allowing to slowly cool.

- Finally, an oxy set can be used to cut metal, including quite thick steel plate. It doesn't give the neat edge of water-jet cutting or the (slightly less neat edge) of plasma cutting, but the steel plate of the turbo exhaust manifold I built a few years ago was oxy cut, the edge being then cleaned-up with a file and grinder.

So an oxy-acetylene kit allows you to do nearly everything required when welding, brazing, softening/hardening or cutting of metals is needed. However, compared with other welding and cutting techniques, it's slower – which is a downside in production work. But for one-offs and home workshop use, the slow pace of the work allows far better user control.

For example, I have recently been brazing together some very thin wall (0.9mm (1/32in)) high tensile (chrome moly) steel tubing. The brazing rods being used are nickel bronze – a very strong brazing material. Normally, welding such thin wall steel tube would be very difficult. (These tubes are quite a lot thinner than exhaust tube,

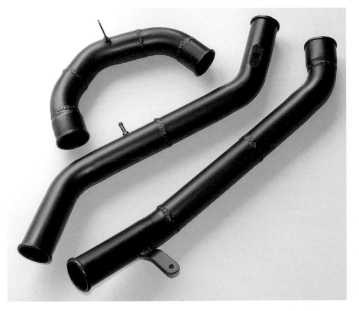

Intercooler plumbing constructed from mild steel, mandrel-bent, exhaust pipe bends that I fusion welded together with an oxy-acetylene kit.

for example.) But brazing these tubes with the oxy kit is child's play. Why? Well, I can braze each joint without fear of melting away the parent material, I can add or remove heat as easily as applying or removing the flame (and of course, also set the starting point by the appropriate selection of flame and tip size), and I can tack the joints and then came back later and seamlessly extend them to full welds. I can also make nuts captive by brazing them in place, and I can easily do tricky things like brazing a disc flush over the end of the tube. And, if I make a mistake, I can very easily 'un-tack' the braze. Now brazing isn't as strong as MIG or fusion welding, but its versatility and ease of control make it unbeatable for my skill level in this application. (Incidentally, at least one well-known book on welding suggests you should never use nickel bronze to braze together chrome moly tube. My experience, and that of many others over the last 50+ years, is that prohibition is false.)

Another point is that an oxy-acetylene kit is best suited for smaller, fiddly jobs. If I were to build a trailer, I'd use electric welding – arc or MIG. You could certainly do it with an oxy, but you'd be there a long time. In the same way, material thicker than about 5mm ($^{13}/_{64}$in) is usually electrically welded. Finally, while brazing rods are available for aluminium, the success or otherwise of this depends a lot on the exact make-up of the aluminium (something normally unknown!). So an oxy kit is not normally used to weld aluminium (although that's how aircraft fuel tanks were done in World War II).

OK, enough of the prelude: let's look at the equipment.

## Equipment

As the name suggests, oxy-acetylene welding uses two gases – acetylene and oxygen. The acetylene is the fuel and the oxygen helps achieve the very high 3100°C (5600°F) flame temperature required.

Note: some details on oxy-acetylene equipment vary from country to country. The following is a generic description – aspects like the colour of the hoses or cylinders, thread types and so on may vary.

Acetylene cylinders are filled with a porous mass which is saturated with acetone. The acetylene dissolves in the acetone much like carbon dioxide is dissolved in the liquid in a soft drink bottle: when the pressure is lowered, the acetylene bubbles out of the acetone. Acetylene cylinders are shorter than oxygen cylinders, are painted maroon (deep red) in colour, and use a left-hand (ie reversed) thread to prevent inadvertent coupling of oxygen fittings.

Oxygen cylinders are taller than acetylene cylinders. They are painted black and use a conventional right-hand thread. The oxygen is compressed to a larger degree than the acetylene, and so the cylinders use heavy walls and are in turn heavy. Oxygen fittings should be kept completely free of grease or oil; should these contaminants meet oxygen, an explosion can occur.

Each cylinder is equipped with a shut-off valve – like a tap in your bathroom, rotate clockwise to close. Fittings connected to both types of cylinder should be specifically design for the application. For example, copper fittings should not be connected to acetylene cylinders as the copper reacts with the acetylene, creating highly explosive copper acetylide.

Each cylinder uses a pressure regulator. These are used to reduce the pressure from the massive bottle pressure to that which is suitable for use. Each regulator has two gauges. One shows the bottle pressure (so giving an indication of how much gas is left in the cylinder) and the other shows the set pressure of the gas being fed to the handpiece. Typically, full acetylene and oxygen cylinders will have pressures of 1800kPa (260psi) and 17500kPa (2500psi) respectively, while the gas pressures for normal fusion welding or brazing are set at 50kPa (7psi).

The name given to the combination of the handpiece, control valves and welding tip is 'blowpipe.' The control valves on the blowpipe allow user-variation of the flow of the two gases. These controls are very important as they allow two things: (a) setting of the flame intensity, and (b) setting of the ratio of oxygen to acetylene. The handpiece is not only the part you hold, but also contains two tubes that feed the gases to the mixer. As its name suggests, the mixer brings the two gases together. Furthermore, the mixer contains some safety devices preventing burning-back of gases through the hoses.

The welding tip is the curved nozzle through which the mixed gases pass. Tips are available in different sizes, varying with both physical size and the diameter of the orifice at the end. (Tips with small orifices are physically smaller overall.) Welding tips are easily swapped as required – they simply unscrew from the blowpipe. Hoses are used to connect the regulators to the blowpipe. These hoses are colour-coded – blue for oxygen and red for acetylene. Finally, flashback arrestors are sometimes fitted to the blowpipe. These lessen the chance of the flame burning its way back towards the cylinders.

## Setting up

Unlike arc welding that I covered earlier, an oxy-acetylene kit has some potentially major safety issues. The bottled gases are under very high pressures, are extremely flammable when mixed, and even when unmixed are hazardous. A hose leaking at a fitting is clearly very dangerous. (Note: hose and regulator fittings should be done up with an appropriate spanner. But don't go mad with tightening torque – a nip-up is sufficient.)

The following process should be followed when initially setting up the gear:

1. Ensure your hands are free of grease and oil.
2. After ensuring that there is no source of ignition in the vicinity (including gas water heater pilot flames!), momentarily open the cylinder valve to blast any foreign bodies from the outlets.
3. Make sure the regulator knob is undone (rotated anti-clockwise until loose) and then attach the regulator to the cylinder. (Remember the threads are different direction, depending on the cylinder.)
4. Open the cylinder valve slowly. The high-pressure gauge will show full cylinder pressure.
5. Check for leakage by closing the cylinder valve and checking that the indicated pressure does not drop.
6. Do the same with the other cylinder.

With the regulators safely on the cylinders, you have now completed the first step – you have low pressure gas available from the cylinders. Now to get that gas to the blowpipe.

1. Connect the appropriate coloured hoses to the appropriate cylinders – blue to oxygen, red to acetylene.
2. Purge the air from the hoses by momentarily screwing down the regulator knobs. You'll be able to hear gas flowing from the end of the hoses. Firstly, ensure that there is no source of ignition in the vicinity.
3. If you are using flashback arrestors (some kits don't include them), connect them to the ends of the hoses. The blue hose goes to the oxygen arrestor and the red hose to the acetylene arrestor.
4. Connect the hoses to the blowpipe. If you are using an off-the-shelf kit, the hoses will be staggered in length to match the offset of the blowpipe fittings.

OK, so now you have the complete system set up and ready for working. Now for a very important test.

1. Close the blowpipe taps and screw in the regulator knobs until each associated pressure gauge indicates 50kPa (7psi).
2. Now close the bottle valves and check that the 50kPa (7psi) gauge readings don't slowly drop. If they drop in reading, there is a leak!
3. If there is no leak, open the bottle valves again.
4. If there is a leak, check all fittings and hoses.

Now we're getting close to being able to weld! Select the appropriate welding tip and screw it into the blowpipe. The accompanying table shows how to go about selecting the tip.

Finally, check that delivery pressure is maintained (ie 50kPa (7psi)) when the blowpipe taps are opened. Again, ensure that there is no source of ignition in the vicinity. If the delivery pressure drops, open the main bottle valves further.

| Plate (mm) | Tip size | Acetylene | Oxygen (kPa) |
|---|---|---|---|
| 0.8 | 6 | | |
| 1.0 | 8 | | |
| 1.6 | 10 | All 50kPa (7psi) | All 50kPa (7psi) |
| 2.5 | 12 | | |
| 3.5 | 15 | | |
| 5.0 | 20 | | |
| 8.0 | 26 | | |

## Shutting down

We haven't even started welding, but it's best at this point to cover shutting the system down.

1. Close the main cylinder valves.
2. Unscrew the regulator handles.
3. Open the blowpipe valves and release the gas in the hoses. Again, ensure that there is no source of ignition in the vicinity.
4. Close the blowpipe valves.

## Flames

Now it's time to light the flame and do some brazing, the easiest of the different oxy welding techniques. Before lighting the flame, you should be wearing welding gloves and welding goggles, the latter with the tinted lenses flipped up. You should also have available a flint lighter designed for lighting oxy-acetylene flames.

Open the acetylene blowpipe valve a little: in a quiet workplace, you should just be able to hear the gas flowing. Hold the nozzle pointing away from you and light the acetylene with the flint lighter. When the gas has ignited, open the acetylene blowpipe valve further until black smoke and soot cease being produced. Next, flip down the tinted lenses of the goggles and slowly open the oxygen valve. The colour and shape of the flame will immediately alter. (If the flame goes out, you've probably opened the oxygen valve too quickly and too far. Turn off the oxygen, turn down the acetylene, and start the process again.)

When you adjust the ratio of oxygen to acetylene you'll soon see that three distinctively different flames can be produced.

A carburising (or reducing) flame occurs when there is excess acetylene. In appearance, the carburising flame has three distinctive parts – the outer flame envelope and two inner cones, where the innermost flame cone is surrounded by a luminous feather. Increasing the amount of oxygen will cause one of the inner cones to disappear. This flame is said to be a neutral flame. The remaining inner cone is long and sharply defined. A flame with an excess of oxygen (an oxidising flame) has a shortened innermost cone.

Not quite a car, but a vehicle nonetheless. I used oxy-acetylene welding to nickel-bronze braze the high tensile, thin-wall tubing to build my recumbent pedal trike. Incidentally, note the air suspension. (Courtesy Georgina Edgar)

In most work, a neutral flame is used. However, an oxidising flame can be used on brass alloys, as the loss of zinc is reduced. A carburising flame is used on steels being hard-faced as the carbon in the excess acetylene is absorbed into the surface of the steel. Always check during use that the flame remains as you have set it.

## Brazing steps

Brazing is the easiest of the metal joining techniques achievable with oxy-acetylene gear. In this type of welding, the brazing rod melts and becomes the glue that sticks the surfaces together. Therefore, the rod must be matched to the application – there's no such thing as a universal rod that will work with all metals. Brazing rods vary in three characteristics:
* thickness
* material
* flux-coated or bare

Let's look at them in turn.

As with electric arc welding, the diameter of the rod should be proportionally matched to the thickness of the material being welded. That is, the thicker the material, the thicker the rod. If the rod is too thick for the application, it will take too long to melt, and, as a result, the materials being welded may melt rather than just getting dull red. A rod that is too thin will melt off before the metals being welded are hot enough, and so the braze won't 'take.'

The material from which the brazing rod is made depends on the application – that is, the materials being brazed together, the required strength, and appearance. There are rods to suit different base metals, different working temperatures and giving different strengths. Bare rods require the addition of a flux. The flux, which cleans

the base materials of surface oxides, is applied to the work either directly by means of a brush or by heating the rod and then dipping it in the flux, so causing the flux to adhere to the rod. Flux-coated rods come with the flux already on the rod.

Here are the steps to brazing. (This material is based on information provided by BrazeTec.)

### Step 1 – Cleaning
Oxide layers and foreign matter such as rust and scales must be removed from the brazing joint either mechanically or chemically before brazing. In the case of sensitive workpieces, thick layers of grease or oil can be wiped off or removed with solvents (eg acetone). Polished workshop pieces do not require any cleaning. Any oxide remaining on the workpiece after precleaning will be dissolved by the flux. If using liquid cleaners, ensure that the surface is dry before proceeding.

### Step 2 – Applying flux
The flux paste is applied to the cold workpiece using a brush. Most fluxes are slightly corrosive and skin contact, particularly with wounds, should be avoided.

### Step 3 – Fixing the workpieces
The pieces to be joined must be fixed in the correct position until the brazing alloy sets. A narrow brazing gap of between 0.05 mm (0.002in) and 0.2 mm (0.008in) is to be set if possible.

### Step 4 – Heating the brazing joint evenly
The brazing gap must be heated evenly to working temperature, so that the brazing alloy can fill the gap. The brazing alloy selected should reach working temperature within three minutes at most. Overheating will damage the braze and the workpiece.

### Step 5 – Placing the brazing alloy on the brazing gap
The brazing alloy can be placed on the brazing gap when the flux has melted to an even glass flow, and the working temperature has been reached. The brazing alloy fills the narrow brazing gap and rises upwards (if necessary) against gravitational force.

### Step 6 – Cooling
When the brazing alloy has filled the brazing gap, the workpiece must be left to cool until the brazing alloy returns to its solid state. The workpiece can then be rinsed in water.

### Step 7 – Removing flux residue
Residual flux must be removed after brazing to prevent corrosion. Where possible, use water or a brush to remove any flux residue (I often use a bead blaster).

## Technique

As with all welding, best results come from practice.

If you are not using flux-coated rods, apply flux to the join. If you are using flux-coated rods, just ensure the join is clean. Use a neutral to slightly oxidising flame. Hold the blowpipe so that the inner cone of the flame is just above the workpiece, and heat the two surfaces until they are a dull red. Introduce the brazing rod and it should melt on application to the metals. The brazing material will then flow into and along the join, following the heat.

Where the gap is very small, just a tiny amount of rod is needed. In this situation, apply the rod and then withdraw it, using the flame to heat the metal ahead of the braze and so make it flow forward into the joint. Where the gap is larger, or a fillet is to be built up, make sure that the brazing material has first taken to the surfaces before applying more rod and building the fillet.

If you get the join too hot, the brazing material will sizzle and spark; too cold and the braze will sit in blobs and not flow into and along the surface. Control is obtained by removing and applying the flame as necessary to maintain the correct heat, and removing and applying the rod as required to add the correct amount of filler material.

The benefits of brazing include reduced heat when compared with welding, so resulting in less distortion of the workpiece. The fact that the parent material is not melted means it's far easier to weld very thin gauges, and – where tight-fitting joins are being brazed – the result is very neat without further work being needed. Brazing is also very easy to do, even by a beginner. The downsides are that the ultimate strength is usually lower than achieved by fusion, arc, MIG or TIG welding (and that's especially the case at very high temperatures), and the brazing process is usually slower than electric welding. Finally, some brazing rods are quite expensive.

I made this unique brass fitting by silver-soldering two existing fittings together. In this type of application, silver soldering is quick, easy and strong. It's part of a custom car air suspension system.

## SILVER SOLDERING

The above section also applies in large part to silver soldering (sometimes called hard soldering). Silver soldering is best used on brass and copper fittings. (Brazing these fittings is also possible, but as the braze melting point is close to the melting point of the parent metals, it's easy to destroy fittings!)

I often silver solder fittings together when I want to create a unique hydraulic or pneumatic fitting. For example, I use thick-walled Sodastream gas cylinders as accumulators (reservoirs) in my car air suspension system. These cylinders come with a unique brass fitting that screws into the cylinder. To allow the connection of a normal hose fitting (eg BSP or NPT), I cut off the end of the Sodastream fitting and silver solder a new brass fitting to the original fitting.

To silver solder, you need specific rods (eg 15 per cent silver) and a flux designed for this purpose. When silver soldering brass, be careful that the joint is tight-fitting and that you don't overheat the brass. Prior to assembly, apply the flux to the fitting's mating surfaces, and also dip the end of the rod in it. Heat the fitting evenly, applying the rod to the workpiece every now and again to test the temperature. When the rod melts easily, apply some additional silver solder and 'chase' it around the join with the flame. Not much silver solder will be needed, and if you're neat, no clean-up will be required except a quick scrub with a wire brush.

Silver solder is expensive but because you use so little of it on good joins, the actual cost per join is quite low – and it is cheaper (and much quicker) than machining a unique brass fitting.

## FUSION WELDING

In fusion welding, you heat the materials sufficiently that they melt together. For example, in a butt joint, you create a pool of molten metal at the join, add a little filler stick (a thin wire rod made of the same material as you're welding) and then move along.

Most beginners, especially those who are 'graduating' from brazing, are reluctant to get the material hot enough. The pool of molten material has to be just that – molten – and the filler rod should melt as soon as you place the end of it in the pool. To make a neat fusion weld, always move in the one direction and do so in a series of small, even steps. Heat the metal until the pool forms, add the filler rod, move forward. It helps if you learn to shuffle the filler rod through your fingers.

With my access to a MIG, I use my oxy for fusion welding very rarely now. In fact, about the only time I do so is when I need to work on a small and fiddly job, and one that is going to be subjected to temperatures too high for brazing to be effective. An example might be welding a pressure-sensing tube to a turbo dump pipe. I also fusion weld steel intercooler plumbing together, mostly because compared with MIG welding, it gives a neat result that doesn't need grinding-back.

## MIG WELDING

If you want to have only one welder in your workshop, I suggest a quality MIG welder with its own shielding gas cylinder.

A MIG welder develops a high current at a low voltage like an arc welder, but instead of using consumable electrodes, a copper-coated steel wire is fed by motorised rollers into the weld. Shielding gas is supplied from a cylinder via a regulator and solenoid. You connect a ground lead to the workpiece, point the nozzle of the handpiece at the work and pull the trigger. The wire is fed out of the handpiece and the shielding gas flows. Once the wire contacts the workpiece the arc starts – no special technique is needed.

People tend not to say this, but compared with all other welding approaches, using a quality MIG is ridiculously easy. How easy? Well, I guarantee that if you are given some basic help by a competent welder, you'll be making decent quality MIG welds in literally five minutes. If you're completely on your own, half an hour of practice will be enough for simple joins like butt welds. Note, though, that I have used only good quality MIG welders that run full shielding gas via a cylinder supply. I cannot say what a really cheap MIG unit will be like, or one that uses flux-cored wire (the wire releases its own shielding gas as its consumed), as I've never used them.

Modern MIG welders typically have only one control knob that sets current. Adjusting this knob also makes the appropriate proportional change to wire speed. Gas pressure is set by a gas regulator to about 20 litres/min (42 cubic feet/hour) – the exception is if you're working in a windy environment that disperses the shielding gas, whereupon you will need to increase flow.

While MIG welders can be set up to weld aluminium, they are usually used to weld steel. A 200-amp welder with adjustable current will weld all steel from 1-10mm (³⁄₆₄ -²⁵⁄₆₄in) in thickness. When set at full power (very rare in my home workshop) the current draw will be high enough to trip the circuit breaker – probably 20A at 240V. In normal use, with materials from 1-5mm (³⁄₆₄-¹³⁄₆₄in) thickness, there are no issues. However, if you live in a country with a 110V supply and/or strict current limiting, investigate this aspect carefully.

Unlike arc welding, even a beginner MIG welder can widen the bead by weaving from side to side, and as there is no slag generated, it's straightforward to place beads on top of one another. By pulling the trigger repeatedly in small bursts, it's even easy to fill holes (holes made for example

A 200-amp inverter-type MIG welder. Add a cylinder of shielding gas, and you'll be able to do nearly all the welding of mild steel that's required in your home workshop. MIG welding is the easiest of the welding types to learn. (Courtesy Toolstop)

by having the welding current too high in error!) and plug and slot welds are also easy. Obviously, some welds are difficult even with a MIG (for example, welding upside down) and safety-critical welding like that on pressure vessels and similar should be undertaken only by experts.

The major beginner error made with MIG welding is to not match the angle of the nozzle with the requirements. The closer the nozzle is to a right-angle to the work, the more the arc is concentrated in one spot. Therefore, it is wise to start off with the nozzle aimed at a right angle to the work, and then once the weld progresses, angle the torch backwards so that the weld is being 'pushed along' in front of the nozzle. This gets sufficient heat into the workpiece to give good initial penetration, and then with that heat present, the weld arc can be spread over a larger area as the weld proceeds. When MIG welding, don't back-track.

It's permissible to weld away from you or towards you, from left to right or from right to left. I suggest that you standardise on one technique. I weld from right to left (I am right-handed) and try to set the workpiece so that I am welding only horizontally. (Of course, a professional welder

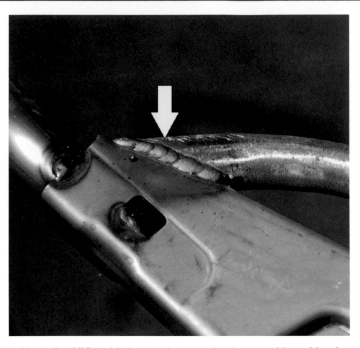

Here, I've MIG welded curved supporting bars to either side of a seat back to provide improved lateral support. (The bars were then covered in foam and seat upholstery.) The arrow points to the weld I made; the other two welds are standard.

can weld from right to left, left to right, up to down, down to up, up and down hills – and so on.)

If the current is too high for the thickness of material, the weld will sink beneath the surface ('undercut'), and may even get sufficiently hot to blow holes. If the current is too low, penetration will be poor, and the weld will be largely just sitting on the surface. If you have to lean in one direction, lean towards too much current rather than not enough (more welds are made that have insufficient penetration than those that have undercutting).

A disadvantage of MIG welding is that, even on low current and with slow wire feed, a MIG still produces a sizeable bead. For example, if welding body panels, exhaust tube or intercooler plumbing, quite a lot of grinding-back will be needed to achieve a flush surface. This is problematic in terms of time, and also because it's easy with these thin materials to grind too far and end up with a weak join. (It's for this reason that I fusion weld these materials with my oxy-acetylene gear.)

If you've only ever struggled with arc welding, I guarantee that you'll be amazed at what you can easily and quickly do with a quality MIG.

## SPOT WELDING
Spot welding, also known as resistance spot welding, is one of the oldest of electrical welding processes. Elihu Thompson, an English-born American engineer, invented

spot welding, with his major patent in this area awarded in 1885. He described this welding as a simple process:

"All that was required was a transformer with a primary to be connected to [high voltage power] and a secondary of a few turns of massive copper cable. The ends of this cable were fitted with strong clamps which grasped the pieces of metal to be welded and forced them tightly together. The heavy current flowing through the joint created such a high heat that the metal was melted and run together."

The transformer comprises a primary winding (connected to mains power) and a secondary winding, connected to the welding tongs or arms. The transformer reduces mains (line) voltage (eg 240V) to a much lower voltage (eg 2V). At the same time, the available current at the welding electrodes increases massively. For example, if the current drawn from the mains socket is 15A, and the voltage step-down ratio is 120 times, then the current available at the welding tongs is 120 times greater, or 1800A! (That assumes a perfectly efficient transformer, but you get the idea.)

When such high current is passed through the sandwich of clamped sheet metals, the higher resistance that is present at the join between the two metals causes heat to be generated. That heat is sufficient to create a 'nugget' of molten material that, when it cools, has joined the two sheets. No metal is added and the weld is actually within the join of the two sheets, rather than external to them.

For the heat to be generated at the junction of the two metals, and not elsewhere in the secondary circuit, all other resistances must be extremely low. It is for this reason that spot welders use thick copper bars to form the arms and the electrodes. This is also the reason that, while aluminium can be spot-welded, the currents required are about twice as high as for welding mild steel. (Aluminium is a much better conductor than steel, and so the heat generated at the junction of the two sheets is lower.)

### Welding parameters
The greater the amount of current flowing through the resistance formed by the two sheets being welded, the greater the heating that occurs. In fact, if the current is doubled, the heating value is multiplied by four times, so small changes in current make more of a difference than you'd first expect.

In addition to the amount of current that is flowing, another important factor is the timing of each step of the process. These steps are:

1. Squeeze time – the length of time the sheets are clamped together by the electrodes prior to current flowing.
2. Weld time – how long the current is flowing, often

The old but powerful spot welder that I use. I added the electronic control box on top that has displays for mains current and voltage, and includes an accurate timer. Not seen is the Solid State Relay (SSR) that now does the switching of the primary current.

measured in cycles (that is, the frequency of the AC waveform – usually 50 or 60 cycles per second).

3. Hold time – the length of time the electrodes stay in contact with the weld after it has been formed.

The pressure with which the metal sheets are held together will affect the resistance that is present at the join. If the sheets are clamped too tightly, resistance will be lower (and therefore so will the heating). But if the force is too low, the electrodes may stick to the work, and expulsion of molten material may also occur. The dent formed by the electrodes should never be more than 25 per cent of the thickness of the combined sheets.

Control of weld time is important, and in some machines, it is, in fact, the only aspect able to be altered by the operator. Too long a weld time and the base metal may get so hot it will boil and metal be expelled from the joint; too short a time and the nugget may not form correctly.

The hold time, when the electrodes are in contact with the weld but current is no longer flowing, is important in determining the rate of cooling of the weld. The pressure applied during the hold time also acts to forge the weld when it is still plastic.

The drawing on the left shows a weld created by a spot welder. Note how the 'nugget' (the melted part) is actually within the sheets where they meet. On the right, in contrast, is a plug weld made by a MIG.

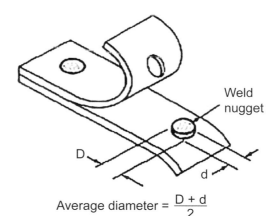

Average diameter = $\dfrac{D + d}{2}$

It's very hard to assess the strength of a spot weld by visual inspection, and so destructive testing of specimens should always be done. This is the peel test – note the nugget size and the fact that it has torn its way through a sheet.

### Testing welds

Unlike a MIG or oxy weld, where external inspection with a trained eye can tell you a lot about the quality of a weld, with spot welding, an external inspection tells you relatively little. Therefore, by far the best way of assessing the quality of a spot weld is by destructive testing a sample.

The two tests are the 'peel' and 'chisel' tests. The peel

test consists of peeling apart a pair of spot welded scraps. A good spot weld is one where:

1.  A hole at least the diameter of the nugget is torn in one of the pieces.
2.  The average diameter of the nugget exceeds that shown in the table below.

| Steel sheet thickness (mm) | Weld nugget diameter (mm) |
|----------------------------|---------------------------|
| 0.6                        | 3.8                       |
| 0.8                        | 4.0                       |
| 0.9                        | 4.3                       |
| 1.2                        | 5.1                       |
| 1.5                        | 5.6                       |

Testing a good quality spot weld made by my old machine.

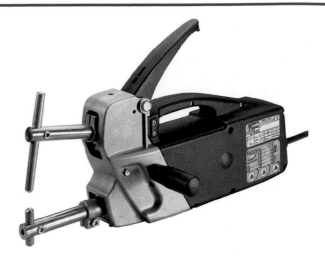

A good quality Telwin spot welder. Beware cheap models that have sharpened electrode ends (giving a tiny welding nugget) and over-stated capabilities. (Courtesy Telwin)

The Telwin Modular 20/TI is typical of the 'hobbyist' breed. It requires a 240V, 15-amp plug, and the maximum short circuit output current is 3800 amps with a secondary voltage of 2V. The unit has a timer adjustable from 0.1 to 1.2 seconds and the clamping force can be varied from 40-120kg (90-260lb). Importantly, the manufacturer suggests that the maximum thickness of steel sheet it can weld is 1mm + 1mm (³⁄₆₄ + ³⁄₆₄in). The Telwin is a 'proper' welder from a reputable manufacturer. So what about the much cheaper units that are available?

Many cheap hand-held spot welders on the market

The chisel test forces a chisel into the gap between adjoining spot welds until either the base metal or the weld fails. During the test, the chisel should not touch the weld nugget. Again, performance of the weld is assessed by looking at the weld nugget diameter and seeing whether a hole is torn in one of the sheets.

### Spot welders in the home workshop

The biggest limitation in the use of spot welders in home workshops is the amount of current that can be drawn from the household supply – and so the amount of current available in the secondary circuit to do the welding. All large industrial welders are 415V three phase, and so can be very powerful. But depending on the country in which you live, single phase, general power points are generally limited to a maximum of 15 amps – and even that usually requires a special circuit from the meter board.

So what single-phase, small hand-held spot welders are available? And what can they weld?

Here I have spot-welded part of a battery tray together. The arrow points to one of the spot welds used. (The battery is now mounted at the rear of the car under the inverted spare wheel).

have electrodes that are tapered down to slim points, in turn making the nugget smaller in diameter. Some cheap units also have less power than the Telwin – and yet claim to be able to weld thicker materials. Obviously, some things here do not add up – you should be quite suspicious of the capability of such welders.

Putting some of the figures in context that have been mentioned in this story, Toyota states that to repair some of their current model cars, a spot welding pressure of 300kg (660lb) force is needed together with a current flow of 10,000 amps for 0.3 seconds. Honda says 350kg (770lb) and 9000 amps. So, putting together a car bodyshell with the same quality spot welds as the factory achieved is not really an option in the home workshop! But that still leaves an enormous range of sheet metal projects that a hand-held spot welder can weld. Any project using raw or galvanised thin sheet metal, where you may have previously used pop-rivets or screws and nuts, can be rapidly spot welded. But it's probably worth saving up for a decent brand name welder, rather than going for a cheapy.

## PLASMA CUTTERS

Home workshop plasma cutters look rather like small welders, but they work on quite different principles.

First, rather than joining materials, they are designed to cut materials. A plasma cutter can be used on a wide range of metals, including brass, cast iron, copper, stainless steel, aluminium and steel. Second, a plasma cutter uses a very high temperature plasma as the cutting tool. (Plasma is the fourth state of matter – the others are solid, liquid and gas.) The plasma is created by heating a gas to beyond a gaseous state, so that it ionises and becomes electrically conductive. This occurs when an electric arc is introduced to the gas (usually air) that is being forced through a small nozzle. The plasma jet can reach over 22,200°C (40,000°F) and blasts away everything conductive that stands in its way. Finally, a plasma cutter uses a higher voltage and lower current than electrical welding, and an additional special high voltage AC supply is used to provide a starting arc to initiate the plasma creation.

Plasma cutting was introduced at the end of the 1950s for cutting metals that could not be cut by an oxy torch. However, the advantages of plasma cutting were so great that it subsequently spread in use to include steels.

Smaller plasma cutters used in home workshops are supplied with compressed air from either an external air compressor, or from a smaller compressor built into the plasma cutter. The handpiece in a smaller plasma cutter is likely to be air-cooled. It contains an electrode tip, a nozzle insulator and a nozzle tip. All these parts are consumables.

In use, a ground cable is connected to the workpiece, the compressor connection is made (if needed), the plasma

A plasma cutter with a 10-25A output, suitable for cutting to a maximum thickness of 9.5mm (3/8in). An external air supply is needed. (Courtesy Lincoln Electric)

cutter is switched on, and the nozzle is held either touching the workpiece (thin materials, eg less than 1½mm or 1/16in) or about 3mm (1/8in) above the workpiece on thicker materials. Press the trigger, and once the plasma discharge has started, move the handpiece along the material.

You must wear gloves and a welding mask or googles, and because a large amount of molten material is sprayed from beneath the cut, also wear full-length leg covering such as a long leather apron. Do the cutting well away from anything that is inflammable. Don't forget that when cutting, the material needs to overhang the welding bench, otherwise you'll tend to cut into the bench top.

While free-hand plasma cutting is possible (eg to cut out a nameplate with a rustic look), most plasma cutting will need the use of a guide. For example, a straight cut is best done with handpiece drawn along a guide made from steel angle that's been clamped into place. Kits are available that allow accurate circles to be cut and provide wheeled holders to keep the cutting nozzle the right distance above the workpiece. Irregular shapes can be duplicated by using a steel template that is first ground to the correct shape and then used as the guide for the handpiece.

I use a small plasma cutter with a built-in compressor; it can have an external air supply added if required. It is rated at 25A with the internal compressor, and 40A with an external air supply. With the external air supply, it can cut up to 9mm (23/64in) steel, 8mm (5/16in) stainless steel and

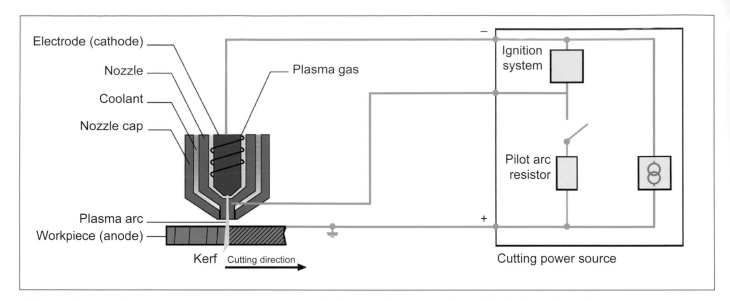

**The creation of a plasma occurs when an electric arc is introduced to the gas (usually air) that is being forced through a small nozzle. The plasma jet can reach a temperature of over 22,200°C (40,000°F). Note the separate power supply for the ignition system that starts the process. (Courtesy BOC)**

4mm ($\frac{5}{32}$in) aluminium – subtract about 25 per cent from these values when using only the internal air compressor.

While it is often suggested that plasma cutting results in a very clean, square edge with no semi-melted material remaining (and that may be true with professional gear), most home workshop plasma cutters will produce an edge that needs some grinding to clean it up. However, the advantage over using, say, thin cutting discs in an angle grinder, is that the cuts are made quickly, and the cuts do not have to be in straight lines. Compared with an oxy being used on steel, heat distortion is also low. Especially when doing major car bodywork repairs or restoration, a plasma cutter will save you a lot of time.

## VISION
The single most important aspect I have found in being able to make good welds is vision. If you have anything but superb natural vision, consider getting spectacles made just for welding. I wear glasses for reading but they don't give the required magnification for welding (or for electronics work, for that matter). To obtain the spectacles I wanted, I asked for some that were appropriate for close-up work – and showed the optician what I meant, holding my hand only about 300mm (12in) from my face. You can get magnifying lenses for welding helmets, but they don't help if you're using oxy-acetylene welding goggles. (You can wear spectacles without issues under all goggles and helmets.) If you find that you wander off the required welding line, or you can't see the welding pool with enough clarity to see if it's shiny or dull, check your vision.

The next step in seeing what you are doing is to get

a really good helmet and/or oxy-acetylene goggles. The goggles are easy – just get a brand-name pair and keep them scrupulously clean. I also wear these goggles when plasma cutting. The welding helmet needs to be a high quality, auto-darkening design with a large viewing area. There are plenty of cheap auto-darkening helmets around, but this is one area where you get what you pay for. Buy a cloth helmet bag for your helmet, and keep it in that when you're not using it.

## A WELDING BENCH
The welding bench that I have been using for years is quite unlike welding benches that I've seen used elsewhere. However, it is cheap, easy to build, flexible in use, and has worked very well for me. It's also an ideal first project with a welder, as the design requires only the fabrication of a simple steel frame.

The bench comprises a square tube steel frame that positions the bench top at a comfortable working height. (When setting this height, remember that some people choose to sit on high stools when they're welding, especially if they're performing a slow welding process.) Across the top of the frame are positioned short supports; these go from the front to the back. I choose to make these supports out of timber, but they could also be made from square steel tube. Sitting on top of the supports are concrete paving stones. This is the working surface of the bench. I use pavers that are 200 x 200 x 40mm (8 x 8 x 1½in). I normally buy more paving stones than I need, and place two or three spares on top as well. If you wish, a shelf can be mounted under the bench top.

**A fold-down welding bench. Dropping the bench reveals a range of stored clamps. (Courtesy Jack Olsen)**

This is a very simple welding bench to make – so what are the advantages? With all the paving stones sitting in place, you have a flat surface that will not be adversely affected by the heat of welding. You can use the spare pavers as supports or weights to hold parts in place. You can also arrange the spares around the workpiece if you're using an oxy-acetylene welder, and want to keep the heat trapped around small items. Dimensionally, pavers are pretty accurate, and so they make easy to use 'squares' when items need to be positioned at 90 degrees to each other. If you are welding a long object, removing a paver or two allows the item to project through the bench-top. This is very useful, for example when you want to weld a cap on the end of a short tube – the tube can project vertically out of the bench top while still being held securely.

Disadvantages? The bench top isn't conductive, so when using an electrical welding or cutting process, you'll need to either connect the ground clamp directly to the workpiece, or use a temporary piece of sheet steel across the top of the bench and connect the earth clamp to that. While I've never had any problems at all, also check that direct heat on the pavers doesn't cause sudden (explosive!) expansion or cracking.

If you're tight for room, a completely different approach is to use a fold-down welding bench. Build a sturdy frame from steel or timber, and place a sheet of steel on top. Use hinges to fasten it to a wall so that when the bench isn't in use, the bench folds up against the wall and takes up very little space. Put your welding gear on wheels so that you can move it to the fold-down table as required. This approach also allows you to store the welding gear out of the way.

## A WELDING TROLLEY

After I bought a plasma cutter, I started running out of floor space – so I decided to make a trolley to stack the plasma cutter above another welder. The trolley had to support a large size gas cylinder and be easily moved around.

I started with an engine stand that I cut and welded into a sturdy rolling chassis. (The engine stand was bought very cheaply at a supermarket that was clearing stock – it was bought just for its parts.) The rear wheels on the trolley were positioned further apart to give overturning stability by providing a wide track at the end carrying the gas cylinder. The gas cylinder sits on two projections from the rear of the frame (they were already there as part of the engine stand), and another part of the engine stand metalwork was used to provide a strong vertical member that the cylinder could be tied to. The cylinder sits in two curved mounts and is held in place by a nylon strap.

A built-up simple frame gives space for the plasma cutter (bottom) and welder (top). This section was made from 40mm (1½in) square tube. Both pieces of equipment just sit on the cross-members – they can't go anywhere, so don't need to be tied down. The verticals extend above the top crossmember far enough to provide convenient points around which the welding cables can be coiled.

In use, the trolley works really well. It rolls easily on its steel castors, and there's a lot more floor space – not only because of the 'stacked' configuration but also because the welding cables aren't on the floor.

This simple-to-build welding trolley mounts a welder and plasma cutter. It was made largely from an engine stand bought at a supermarket clearance. Note the large castors that easily handle the weight.

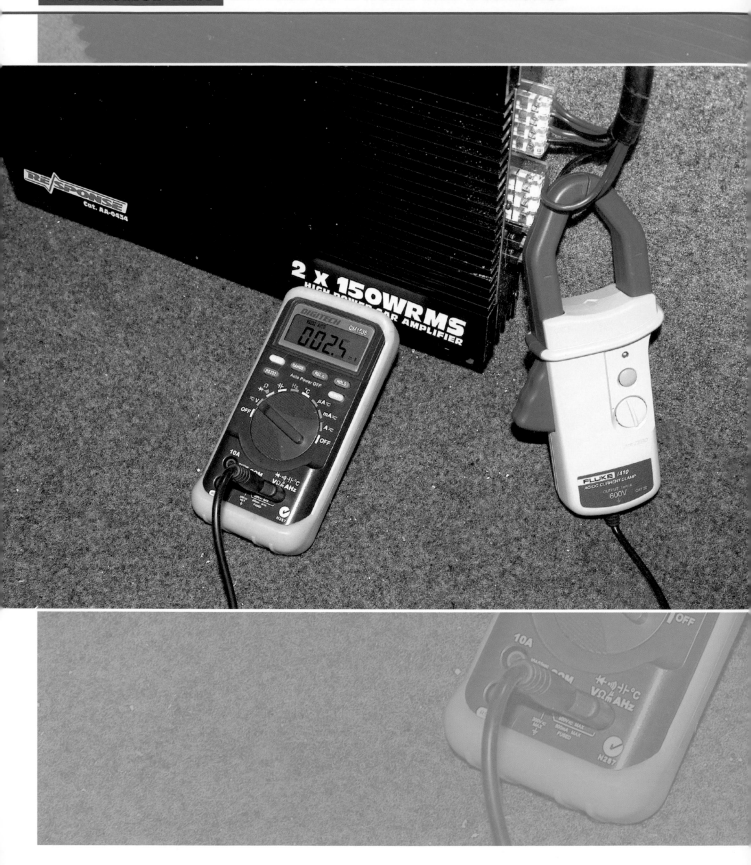

**Chapter 9**

# Tools for car electronics

To perform electrical or electronic work on your car, you'll need appropriate tools and supplies to do it. That should include the basics – wire strippers, side (diagonal) cutters, good quality tape, heat shrink and so on. Here I'll assume that you already have those items, and concentrate on the more complex tools.

## MULTIMETER
### Features
The key piece of equipment is a multimeter. A multimeter is a test tool which can measure a variety of different electrical factors – at minimum: volts, ohms and amps. In addition to these measurements, it is also useful if the meter can display:
* continuity
* frequency
* duty cycle
* temperature (via a plug-in probe)

A meter with max/min/average functions is even more useful. So what are these extra features, and why are they important?

A continuity function sounds a beeper when the meter's probes are connected together. If the probes are connected to either end of a wire, the sounding of the beeper shows that there is no break in the wire – that is, it has continuity. In working on cars, the continuity function is used frequently – to see if the filament in an incandescent bulb has burnt out, to ascertain if a fuse is blown, or to work out which wire is which in a wiring loom.

Frequency refers to how often something is changing – in a car, that usually means being switched on and off. For example, many flow control valves are rapidly pulsed – an idle speed control valve, or a turbo boost control valve. Being able to measure frequency is important if the valves are to be controlled by a new system (for example, programmable engine management) or you are fault-finding.

Duty cycle refers to the proportion of time a device is switched on for. For example, a fuel injector at idle speed might have a duty cycle of only 2 per cent – the injector is open and flowing fuel for only 2 per cent of the time. Especially in a modified car where power has been increased, being able to read the maximum duty cycle is important – at 100 per cent duty cycle, the injectors are working at full capacity. Duty cycle measurements are also used when fault-finding, for example of an idle control valve.

Temperature measurement is via a K-type thermocouple that can be plugged into the meter. Thermocouples are typically available as bead designs (these very small and lightweight which allows their temperature to change very rapidly, but they are fragile) or probe types (much more rugged but their temperature changes more slowly). Often, it's worth having one of each sensor type – bead designs are easy to slip under hose clamps, for example. You can use such a probe to measure the temperature of the coolant, engine and gearbox oil, and intake air.

Many meters have a 'peak hold' or similar function. This displays the maximum reading that occurred during the measuring period. This is especially useful if you are testing on the road and cannot safely watch the meter. For example, for best performance, an engine should breathe cold air. If you are making a new intake, you can use a temperature probe and a multimeter to find which areas in the engine bay get hot – and which stay cool. Moving the

A cheap multimeter – it measures voltage, resistance and current to 10A. Additional functions include temperature (via an additional probe) and continuity. (Courtesy Toolstop)

A more sophisticated (and expensive!) Fluke 189 multimeter that adds frequency and duty cycle measurement, and has high speed minimum, maximum and averaging functions. I use this meter.

probe around and using the peak hold function on the meter will soon give you this information.

More expensive meters not only show the maximum reading that was gained but also can show the minimum and average readings as well. If the meter can sample very rapidly, these additional functions are very useful. For example, I used the output of a height sensor to measure the effectiveness of a front spoiler on a car. If the spoiler was to be effective, you'd expect that ride height would decrease (or at least, not get higher!) as the car went faster. However, on a bumpy road, the output of the sensor is constantly changing, so it's hard to make sense of the reading. What is wanted in this situation is a reading of average height. This allows the ride height to be compared at the same speed with and without the spoiler fitted. To make this measurement, I used my high-speed averaging multimeter.

Multimeters must have what is called a 'high input impedance.' This means that when you apply the meter to the system that you are measuring, the meter won't draw more than a tiny amount of current. Meters that don't have a high input impedance (old analog meters and some cheap digital meters) will load down the system. For example, measuring the output of a narrow-band oxygen sensor will be impossible with a low impedance multimeter – and attempting to do so may well damage the sensor. When looking at meter specs, the meter should have an input impedance of at least 10 mega-ohms.

Different multimeters have a different number of digits on their LCDs. This can be readily seen by looking at the catalogue picture or at the meter in the flesh. But what you see may be not what you're actually getting – there's a trick involved in understanding what the meter can actually show you.

A typical low-cost multimeter has what is called a '1999 count.' That is, it has four digits with the last three digits able to display all numbers from 0-9, but the first digit able to be only 0 (sometimes blanked) or '1.' The highest number that can therefore be displayed is 1999 – or 1.999, 19.99, 199.9. Confusingly, this type of display is often also called a '3½ digit' display – the '½' indicating that the first digit is capable of showing only '1' or '0.' Next up the sophistication list are '3999 count' or '3¾ digit' designs. These have a maximum display number of 3999, 3.999, 39.99, 399.9. Really top meters go as high as '50,000 count' or '4½ digits' and can display numbers like 50,000, 5.0000, 50.000, 500.00, 5000.0. It's easy to get lost in all of this, but remember, the higher the 'count' or 'digit' number, the more detail you can read.

Multimeters are available in auto-ranging or manual-range types. An auto-ranging meter has much fewer selection positions on its main knob – just amps, volts, ohms and temperature, for example. When the probes of the meter are connected to whatever is being measured, the meter will automatically select the right range to show the measurement.

Meters with manual selection must be set to the right range first. On a manual meter, the 'Volts' settings might include 200mV, 2V, 20V, 200V and 500V. When measuring 12V battery voltage in a car, the correct setting would be '20V,' with anything up to 20 volts then able to be measured.

While an auto-ranging meter looks much simpler to use – just set the knob to 'Volts' and the meter does the rest – the meter can be slower to read the measured value. This is because it first needs to work out what range to operate in. If the number dances around for a long time before settling on the right one it can be difficult to make quick measurements, and even more difficult if the factor being measured is changing at the same time as well! However, to speed up readings, some auto-ranging meters also allow you the option of fixing the range. Note that expensive multimeters will very quickly get the right reading, even if they are auto-ranging.

A backlight function is very useful when working with cars – it allows night on-road testing and also makes things easier when working in darkened footwells.

Some meters have two displays, although they still have only one pair of input leads. The two displays are used to simultaneously show two characteristics of the one signal that's being measured. To do this, the two different signals have to be on the same input – you can't show temperature and voltage for example. But if you are measuring (say) a pulsed solenoid that controls turbo boost, you can simultaneously measure both its duty cycle and frequency, and these are shown separately on the two displays. But in most car applications this isn't all that advantageous – you usually only want one parameter measured at a time.

For automotive use, look for a meter design which comes in a brightly-coloured rubber holster – it helps protects the meter from damage as well as making it easier to find – and one which is protected against the entrance of moisture. Good meters use 'O'-rings to seal the case and jacks.

## Using a multimeter

When measuring all but current flow (amps), the multimeter is applied in *parallel* with the circuit. For example, if you want to measure the voltage at a headlight, the negative probe of the meter is grounded (ie connected to body chassis) and the other side is connected to the power supply at the headlight. The meter is set to 'Volts DC' to make this measurement. In nearly all voltage measurements on a car, the negative probe of the meter is grounded and the other probe connects to the signal of interest.

If you want to measure the resistance of a coolant temperature sensor, the sensor is unplugged from the wiring loom, and then the multimeter probes connected to the

An accessory multimeter probe set is a worthwhile buy. It will allow you to easily probe through the insulation of wiring looms, back-probe connectors and connect to car ground terminals.

two terminals. The meter must be set to the correct ohms range. When measuring resistance, always disconnect the component from its circuit.

When measuring current, the meter is inserted into the circuit in *series*, so that the current flows through the meter. You will need to swap the positive lead of the meter to a new socket, and then select the correct range of amps on the dial. For example, if you wished to measure the current flowing to a light, you would need to break the circuit to the light and insert the meter. Often, pulling a fuse and then probing to each side of the fuse socket is the easiest way of doing this. Note that most multimeters are limited in maximum current to a short-term 10A – that's around 140W when the car is running. If you wish to measure higher currents, use a current clamp accessory (covered below).

Ensure that when you have finished measuring current, you return the positive lead of the multimeter back to its general-purpose socket position and de-select amps on the meter dial. If you don't do these things, next time you try to use the meter to measure voltage, you will blow a fuse in the meter or damage it. I've been using multimeters for many years, and this is still an error I occasionally make! And multimeter fuses are expensive …

## Leads

Multimeters come with leads that are equipped with sharp probes. These are fine for general purpose measurements,

but for best car use, you should buy some additional accessory leads. I suggest that you buy extra leads equipped with the following:

- Alligator (crocodile) clips – they are useful when one side of the multimeter needs to be grounded, eg by connecting to a chassis bolt.
- Very sharp insulating-piecing probes – these allow you to tap into a wire without it being disconnected from the circuit. These probes are easily damaged, so take good care of them.
- Spring hook probes that use miniature hooks that will lock around terminals, eg the terminals inside an injector socket, allowing the measurement of injector resistance.

Buy leads that are silicone insulated as they'll be more durable than leads with conventional insulation.

## Current clamp

A current clamp allows you to measure much higher current flows than a normal multimeter can handle. A current clamp outputs a precise voltage per measured amp. For example, it might have an output of 1 millivolt per amp. This makes measuring the clamp's output easy – if the multimeter shows a measurement of 5mV on its voltage scale when connected to the operating clamp, the current flowing in the wire is 5A. If the voltage displayed on the multimeter is 100mV, the current flowing in the wire is 100A.

When using a current clamp, its jaws are opened, the clamp passed over the wire, and the jaws closed. The wire is then centred in the opening and the measurement made. Note that it's the *individual conductor* that is measured – not a cable containing both earth and power leads, for example.

Current clamps are not particularly good at accurately measuring very small currents. This is so for two reasons – firstly, if the output scale of the clamp is 1 millivolt per amp,

A current clamp being used to monitor the current draw of an amplifier. The meter is reading 2.5mV which indicates a draw of 2.5 amps.

a current flow of 0.5A is only 0.5mV – a figure that is getting very low for many multimeters to accurately measure. In addition, because of the influence of stray magnetic fields, current clamps need to be zero'd before they can be used. That is, a knob on the clamp first needs to be turned until the current reading is zero – obviously, when there isn't any current flowing through a wire inside the jaws! Getting the meter to read precisely zero can be fiddly. For these reasons, current clamps are usually used for current measurements of about 5A and upwards. (Most multimeters have a maximum current rating of 10A, so the overlap between a current clamp and a multimeter works fine.)

I use my current clamp when assessing the draw of high-current devices like air suspension compressors, car sound amplifiers and starter motors. I've also used my clamp when measuring alternator output.

## Pressure sensor

Pressure sensors are available that will plug into a multimeter. The sensor can be used to measure fuel pressure, intake manifold pressure (both positive and negative), oil pressure and so on. The benefit of using an electronic sensor over a mechanical pressure gauge is that, depending on the meter being used, you may be able to measure not only the 'live' value but also the maximum, minimum and average values. A fast-response pressure sensor can also be used with an oscilloscope – more on scopes in a moment. Like current clamps, pressure sensors output a certain voltage per unit of pressure. The Fluke PV350 sensor I have can be switched to either metric or Imperial units, and outputs a voltage of 1mV DC per unit.

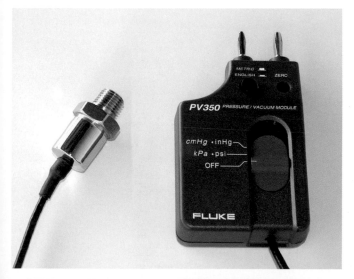

A Fluke PV350 sensor that plugs into a multimeter. It can be switched to provide an output in either metric or Imperial units, and outputs a voltage of 1mV DC per unit.

## Buying a multimeter

If you are new to car electronics, buy a low-cost multimeter – for example one that covers just voltage, current and resistance measurements, and has a continuity function. However, if you want to be able to use functions of the sort mentioned above (max, min, average – and some meters have logging) then you'll be up for a much bigger outlay. In that case, I suggest that you buy a good quality meter. I use a Fluke multimeter for my main measurements and a second, cheaper multimeter when I want to monitor two circuits simultaneously. I also have a Fluke current clamp, pressure sensor and thermometer adapter. However, I don't buy Fluke probes and leads – they are much too expensive for what you get.

## SOLDERING IRON

Next on the equipment list is a soldering iron. A general-purpose mains-powered iron (eg a 25 or 40W design) will perform most soldering tasks on a car – and also can be used to assemble electronic kits and work with individual components. If you can stretch for the extra

A low cost, fixed-power soldering iron. This sort of iron is fine for general automotive soldering. (Courtesy Toolstop)

money, a temperature-controlled iron is a better proposition than a basic iron. Not only do they allow you to dial up an appropriate temp for the job, but they are usually configured with a base station and a remote iron that's linked to it with a supple cord. The lighter iron makes it a lot easier to use, and you get an inbuilt stand and a tip-cleaning sponge.

In addition to the general-purpose iron, a high-powered iron is useful, especially if you are working with heavy-gauge cable like that used with car sound amplifiers, air suspension compressors, alternators and starters. An alternative to buying a heavy-duty soldering iron for these cables is to buy a large crimping tool. Crimping (covered in a moment) has the advantage that the cable remains flexible right up to the lug, whereas soldering tends to make the cable stiffer and brittle.

I have two soldering irons – a 40W variable temperature station and a huge old 250W iron for heavy cables.

When soldering, there are two different techniques that can be taken. Let's take the example of soldering two wires together. The insulation is stripped and then the strands of each cable twisted. Then you can either: (1) tin the two exposed wires (that is, melt solder onto them), then place the wires adjacent to one another, applying heat and a little more solder until the join is made; or (2) twist the un-tinned wires together, and then apply solder and heat until the join is made.

In both cases, the wire must be heated sufficiently that

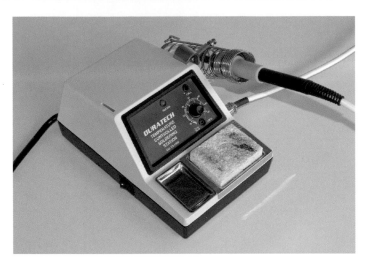

A variable temperature soldering iron. This is a good soldering iron for car and general electronics use.

when solder is applied, it melts and flows easily into the wire strands of the cable. Do not apply solder to the iron's tip and then apply this to the cable – instead, apply the tip of the iron to the cables and then apply solder to them. In many cases, the soldering iron will need to be successively applied to both the front and the back of the join – the solder should have thoroughly 'wetted' the join, and the resulting soldered join should be shiny in appearance. Always use solder designed for electronics work.

## VARIABLE POWER SUPPLY
A variable supply is an absolute must-have. It's especially

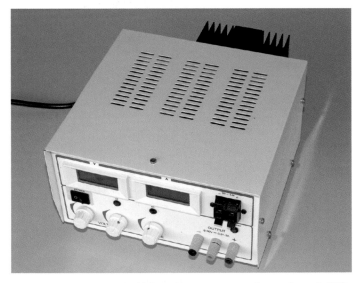

A bench power supply. This unit can output voltages from 0-30V at currents of up to 3 amps. It has digital displays of both current and voltage, and both can be limited in output. More compact units with similar specs are now available.

useful if you want to bench-test an item before it is installed in the car.

A variable power supply can output adjustable voltages, and, in some cases, variably limit the current flow as well. At low prices, you're likely to find power supplies that can supply currents up to 3A and have an output voltage range of 0-20V. However, the higher the current capability, the better. For example, if you wish to test a headlight, you'll typically need a power supply capable of 4 or 5A.

Most important is to have a power supply that has built-in digital meters to show the current being drawn and the voltage being supplied. Having this information instantly available (you can measure it with your multimeter – but then that ties up the meter) is extremely useful. Also look for fine and coarse supply voltage adjustments (so much

A power supply I constructed using readily available modules. At 12V it can output 8 amps for short-term use, and continuously supply 6 amps.

easier to accurately set the voltage), and the previously-mentioned variable current output facility. The power supply should also be completely protected against overload, high temperatures, and short circuits. Compact, switch-mode power supplies take up less bench space than traditional transformer-based designs.

I have several power supplies, with two of them being used frequently. My bench supply is a compact 4A unit that runs to 30V. I also have a self-built 6A, 24V supply. This one can put out 8A for short-term use.

## INFRARED THERMOMETERS
It's not actually a tool for electronics use but I've included it here because it seems to fit – an infrared thermometer.

Infrared thermometers are remote sensing designs. That is, you point the instrument at the surface to be measured and it reads off temperature on a digital display, without any contact having to be made. This is the major advantage

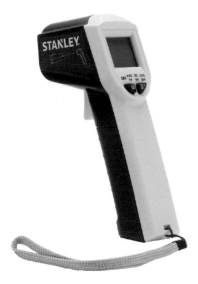

This infrared thermometer will measure temperatures from -38 to 520° C (-36 to 970° F). However, it has a fixed emissivity of 0.95 and so will give incorrect readings for shiny and/or silver surfaces. (Courtesy Toolstop)

Uses for an infrared thermometer on a car include:
- brake temperatures, to see if brakes are dragging and to determine their work share
- engine sump temperatures (eg to see if installing an undertray increases oil temps)
- temperature drop across heat exchangers – engine oil cooler, trans oil cooler, intercooler, radiator
- power steering fluid temperature
- tyre temperatures (eg to assess handling set-up)
- car sound amplifier temperatures

When selecting an infrared thermometer, look at the following specifications.
- Range – many infrared thermometers are limited in their response to high temperatures of the sort likely to be experienced with exhausts and brakes. However, measuring these temperatures in car applications is fairly rare – but you still want a max temperature reading of at least 300°C (570°F).
- Accuracy – this is most affected by the ability to input different emissivities into the instrument. (Note: if the thermometer uses a fixed emissivity (eg 0.95) then any quoted accuracy is likely to be unobtainable when the thermometer is being used for practical purposes.)
- Emissivity – to be able to accurately measure the temperature of a variety of surfaces, the thermometer must have the capability of being programmed with different emissivities.
- Field of view – a narrow field of view (eg 8:1) allows pinpoint temp reading.
- Sighting – how the instrument indicates the area being measured, eg by laser or LED beams
- Data hold – this is important because as soon as you move the thermometer aim away from the sample area, the reading will change. In some situations, reading the

over a traditional thermocouple – you don't need to touch the surface to make the measurement. This improves safety (you won't get your fingers burnt, and the area being measured can be moving or difficult to safely access), and as you don't need to wait for a probe to come up to temperature, the speed with which measurements can be made is high.

An infrared thermometer measures the amount of infrared energy given off by the object to be measured. For a given temperature, the amount of infrared energy depends on what is called the body's emissivity. The emissivity of a perfect radiator of infrared energy, called a blackbody, is 1. However, many objects have emissivities that are less than 1, and if correction isn't made for this change in emissivity, the temperature measurement will be wrong. If the object either (a) reflects infrared energy, or (b) transmits infrared energy, the emissivity value won't equal 1. Shiny polished surfaces, such as aluminium, are so reflective of infrared energy that accurate infrared temp measurements may not be possible. Many infrared thermometers with fixed emissivity use a default value of 0.95. Unoxidised aluminium has an emissivity of just 0.02 at 25°C, while rubber's emissivity is 0.94. So the measurement you just made of your polished intercooler tank is likely to be way out in accuracy, but the tyre temp measurement will be just fine!

To accurately measure the temperature of a variety of surfaces, the infrared thermometer needs to be able to be programmed with the emissivity of the surface that you are working with. If the thermometer doesn't have this capability, look for a better one, or be aware that many readings (especially of shiny and light-coloured surfaces) are likely to be wrong. (If you are working with a thermometer with fixed emissivity and you are trying to measure the temperature of a silver or shiny surface, put a dob of black paint on the surface.)

Measuring the temperature of the brake disc gives a good indication of whether the brakes are dragging. This infrared thermometer is programmable for different emissivities.

display at the same time as it is pointed at the correct area can be difficult. A data hold function overcomes this problem.

I use an SP Tools infrared thermometer for general use, and for really accurate readings, I have a Steinel unit that has programmable emissivities.

## OSCILLOSCOPES

An oscilloscope pictorially shows changing voltage over time, drawing a trace that accurately depicts the pattern of voltage variation. It is your window into the shape of the signals, whereas a multimeter shows you just the magnitude of the parameter. A scope is the only way that you're going to be able to look at signals coming out of camshaft and crankshaft position sensors, speed sensors and ABS sensors, among others. And it's also the only way that you're going to be able to see the signals going to injectors, idle air control valves, boost control solenoids, auto trans pressure control solenoids, and so on.

Traditionally, scopes have been used by mechanics to look at primary (low voltage) and secondary (high voltage) ignition signals. And that's a valuable use for a scope. But these days a scope is far more likely to be used to look at inputs and outputs of an Electronic Control Unit (ECU). In fact, most good factory workshop manuals now show sample scope traces, so that you can use a scope to quickly find if the output signal from the sensor or ECU looks as it should.

As suggested earlier, an oscilloscope is basically a graph-displaying device – it draws a graph of an electrical signal. In all automotive applications, the graph shows how signals change over time: the vertical (Y) axis represents voltage, and the horizontal (X) axis represents time. This graph can tell you many things about a signal, such as:

- the time and voltage values of a signal (how many volts and when it changes)
- the frequency of an oscillating signal (how often the voltage is rising and falling)
- the frequency with which a particular portion of the signal is occurring relative to other portions (is there a part of the signal that varies more rapidly up and down than other parts?)
- whether or not a malfunctioning component is distorting the signal (do the sine waves look more like square waves?)
- how much of the signal is noise and whether the noise is changing with time ('noise' is normally seen as a superimposed signal – jagged edges on a sine wave, for example)

Oscilloscopes can be classified as analog and digital types.

### Analog oscilloscopes

An analog oscilloscope works by applying the measured signal voltage directly to the vertical axis of an electron beam that moves from left to right across the oscilloscope screen – usually a cathode-ray tube (CRT). The back side of the screen is treated with luminous phosphor that glows wherever the electron beam hits it. The signal voltage deflects the beam up and down proportionally as it moves horizontally across the display, tracing the waveform on the screen.

Analog oscilloscopes are characterised by the large screens used in traditional 'tune-up' machines and the smaller older scopes with the glowing green screens used in electronics work. They are excellent tools, however in automotive use they suffer from major drawbacks – the need for mains (household) power, the greater difficulty in set-up, and the absence of a storage mode that allows the freezing of the on-screen image.

### Digital oscilloscopes

A digital oscilloscope uses an analog-to-digital converter (ADC) to convert the measured voltage into digital information. It acquires the waveform as a series of samples, and stores these samples until it accumulates enough samples to describe a waveform. It then re-assembles the waveform for display on the screen.

The digital approach means that the oscilloscope can display any frequency within its range with stability, brightness, and clarity. It can also easily freeze the waveform, allowing it to be studied at leisure. Digital scopes can usually be powered by batteries and use an LCD screen. All scope adaptors that are used with laptop PCs are digital. Digital scopes will usually also calculate the frequency and duty cycle of the signal that you are monitoring.

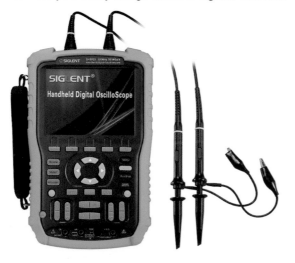

**I use this colour Siglent SHS806 two-channel oscilloscope. This unit can also perform multimeter and logging functions. (Courtesy Siglent)**

Oscilloscope modules are available to turn a laptop PC into a fully-fledged digital oscilloscope. This scope can monitor four separate channels simultaneously, and is powered from the laptop's USB port. (Courtesy Toolstop)

## Digital scope specifications

As briefly indicated above, an analog scope effectively draws the waveform as it occurs. However, a digital scope samples the voltages coming into the scope – it isn't continuously measuring the input signal, but instead is measuring only bits of it. The waveform is then reconstructed from these separate samples and displayed on the screen. It's a join-the-dots process. How often the scope samples the signal is known as sampling speed, expressed in samples/second. All else being equal, the

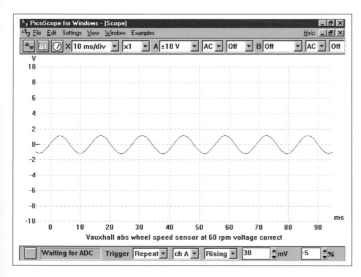

This scope trace shows the output of an ABS sensor. As can be seen, it is an alternating current (AC) waveform. (Courtesy PicoScope)

higher the sampling speed, the higher the frequency of the signal that can be accurately displayed. Or to put it another way, the higher the sampling speed, the shorter the event that can be captured. In addition to sampling speed, the maximum frequency that the scope can accurately measure is also influenced by the scope's input amplifiers and filters. This factor is called 'bandwidth.'

The amount of memory that the scope has is also important – especially in automotive applications. Memory (sometimes referred to as Record Length or Buffer Size) is relevant for two reasons.

* The more closely spaced the samples are, the more memory that's required to hold them before a complete waveform can be displayed. In other words, high sampling rates require more memory.
* The longer the length of time over which the waveform needs to be displayed (called the time-base), the more samples that need to be kept if the sampling resolution is to be retained.

In automotive use, where most often quite slow time-bases are used, the second point is the more important of the two. For example, when the full width of the screen shows 90 nano-seconds, a scope may be able to sample at 10 mega-samples/second, but if you lengthen the period that you want to display to 9 milliseconds, the effective sampling rate (dictated by how many samples can be memorised) may drop to only 10 kilo-samples/second. In some scopes you can go even further, setting the time-base to hours! In this case you want lots and lots of memory if you're to be able to store what's basically become a data-log of the signal.

The amount of memory available is also relevant if the scope has the ability to zoom in on waveforms after they have been frozen. In order to gain that extra waveform detail, more memory will be required, especially if you have a long time-base.

In addition to sampling speed and bandwidth, the analog to digital converter (ADC) resolution of the scope is important. Most have 8-bit vertical resolution which limits the voltage variation that can be measured to just under 0.4 per cent. On the other hand, 12-bit scopes can resolve changes in voltage levels of only 0.024 per cent.

Especially in designs where add-on modules are used to turn laptop PCs into digital scopes, the functionality of the software is important. In addition to scope functions, many of these designs can also act as spectrum analysers (that is, showing on a vertical bar graph the magnitudes of all the different frequencies), multimeters (although often with quite limited ranges), and as data-loggers. All manufacturers of this type of scope allow web downloads of demo, trial or fully functioning versions of the software, so you can play before you buy the associated hardware.

Even the cheapest handheld digital scope will show

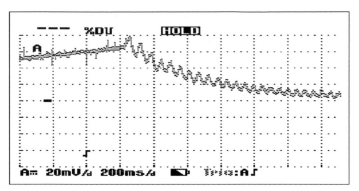

This is a scope trace of manifold pressure, measured between the throttle body and the turbo using an electronic pressure sensor and a Fluke 123 Scopemeter. The throttle was closed abruptly when the engine was on boost. Without a blow-off valve, the pressure waves of up to 20kPa (nearly 3psi) can be clearly seen racing up and down the intake system between the throttle and the turbo. Electronic sensing and good measuring equipment allows you to see fascinating things!

you the waveform of most car input and output signals – so you'll immediately be ahead of what you'd be seeing with a multimeter. However, as you go higher in specs – especially in sampling rate and bandwidth – you can be more confident of seeing a better representation of the original waveforms.

A scope is used in largely the same way as I described earlier for a multimeter. The earth probe of the scope (usually in the form of a crocodile clip) is connected to the car chassis, and the other probe is connected to the signal being monitored. Most digital scopes have an 'auto' button that when pressed, configures the horizontal (time-base) and vertical (voltage) axes to give a good view of the signal.

### OBD READERS

Readers that can display the output of the Onboard Diagnostics Port (OBD) allow easy fault-finding and resetting of fault codes. An OBD output is available in vehicles produced after 1996 (USA) and 2001 (Europe). The standard OBD parameters include, among many others:
- fuel system status
- calculated load value
- engine coolant temperature
- short term fuel trim
- long-term fuel trim
- fuel pressure
- intake manifold absolute pressure
- engine rpm
- vehicle speed sensor
- ignition timing advance for #1 cylinder
- intake air temperature

Generic OBD readers are very cheap, but be wary of putting down your cash without first investigating the functionality of the tool. In addition to those parameters provided under the

standards, individual manufacturers use their own extended data read-outs and fault codes. To be most effective, the code reader you use needs to be compatible with as many of the manufacturer-specific codes for your car as possible. In addition, you want to be able to access all the different electronic car systems – and there may be a dozen of those, from ABS to airbags to transmission control.

It's impossible to generalise about which reader will be most suitable for your application. The best approach is to go to an online discussion group that specialises in your make and model of car and ask the people there what they use. In some cases, enthusiasts or private companies have developed new hardware and software to best display operating parameters and fault codes for specific cars and/or manufacturers. Some of these packages can even achieve what is normally a dealer-only processes – changing the action of convenience features, or coding new keys or modules. For example, VCDS for Volkswagen/Audi cars is, as the seller says, "a software package for Windows that emulates the functions of the dealers' very expensive proprietary scan tools." When I owned a Volkswagen product, I used VCDS.

Generic OBD displays are available as colour dashboard displays, including head-up displays that show the readings reflected on the inside of the windscreen. However, such displays are less useful as diagnostic tools because they typically display only a quite limited number of parameters. While not as flashy as some of the readers that are available, ScanGauge has a good reputation as a general purpose OBD reader and display, and can clear fault codes. I have used a ScanGauge and it worked just as advertised.

If you're working on a car that has engine management but pre-dates OBD, you will likely be able to trigger a mode that displays fault codes by the flashing the Check Engine Light (CEL), or in even older cars, LEDs mounted on the engine management control unit itself. Of course, if you're dealing with a carby-and-points car, get out the multimeter!

In not quite the same category as OBD readers are tools that can be used to reset service indicators – for example oil change warnings. Some of these tools, designed for use with only one make and model, are very cheap.

The ScanGauge connects to the OBD port. It has a good reputation as an in-cabin display that can read and cancel fault codes. (Courtesy ScanGauge)

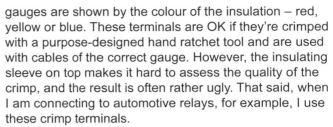

**A generic OBD reader that can also reset oil service/inspection lights.
(Courtesy Toolstop)**

OBD readers simply plug into the OBD socket – no other installation is needed.

## CRIMPING PLIERS

For many years, I soldered all of my wiring connections on cars. And despite many people telling me that such joints were doomed to failure as the brittle wires broke near the join due to vibration, I can honestly say that I have never had even one join fail in this way. (Always use heat-shrink or good quality electrical tape around soldered joins, and cable-tie wiring so that it cannot move. Doing this gives soldered joints much greater reliability.) However, when

I installed programmable engine management in one of my cars, I had to change my approach – the wiring loom plugs all required specific crimping to attach the loom to the pins.

Many people's experience with crimped car terminals involves the use of pre-insulated crimp terminals widely sold in automotive parts shops for car wiring. These terminals are available in male and female spade, ring, and bullet types. Different wire

**This pair of ratchet crimping pliers is suitable only for the common 'red-blue-yellow' insulated terminals often used in car wiring. If you intend to also work with engine management and other car system connectors, you'll need pliers that come with a range of crimping dies. (Courtesy Toolstop)**

gauges are shown by the colour of the insulation – red, yellow or blue. These terminals are OK if they're crimped with a purpose-designed hand ratchet tool and are used with cables of the correct gauge. However, the insulating sleeve on top makes it hard to assess the quality of the crimp, and the result is often rather ugly. That said, when I am connecting to automotive relays, for example, I use these crimp terminals.

High quality plugs and sockets use specialised, bare crimping terminals. My favourite plugs and sockets are Deutsch DT. These are available in a variety of pin numbers, and are waterproof, so suitable for mounting under the car or in the engine bay. They also use a crimping system that works very well. You'll need to buy the correct crimping tool – kits are available that incorporate a crimper and a variety of plugs and sockets. The crimping tool completely surrounds the sleeve-like terminal and the resulting crimps are very secure – with little crimping effort or time needed. In contrast, some engine management system plugs use crimp terminals that need expensive tools – or, alternatively, cheaper tools where the crimp has to be made in a number of steps.

If you need a general-purpose crimping tool, buy a ratchet design that has interchangeable dies supplied with the tool. With these you should be able to get competent results with a wide variety of terminals – although not the Deutsch DT terminals that are unique and so need their own tool. Unlike the plugs and sockets, the crimping terminals themselves are cheap, and so it's wise to practice with a scrap piece of wire until you get a strong crimped connection that won't pull off with a sharp tug.

If you wish to crimp heavy-duty lugs for battery, starter and alternator cables, you'll need a large crimping tool. These look a bit like bolt-cutters (they have long handles) and feature rotary heads that allow a variety of terminal sizes to be crimped.

**A full kit of Deutsch DT plugs and sockets, complete with the special crimping pliers (right). If you do a lot of car wiring, these plugs and sockets are unbeatable. (Courtesy Retrofitparts)**

**Chapter 10**

# Organising a home workshop

A key aspect of having an effective and productive home workshop is its organisation. Where you put things and how well they work together can make or break or a workshop. But of course, not everyone has the same available space, equipment or budget, so let's look at a wide variety of home automotive workshops.

## MINI WORKSHOPS

If you're in a position where you have no real working area for your car, what do you do? One approach is to use a mini-workshop, like a wooden chest that folds up when not in use. For example, if you like doing electronic projects on your car, you can position the electronics workbench temporarily on your kitchen table, complete the project, and then take the project to the car to be fitted.

Morten Nisker Toppenberg lives in Denmark, and has constructed the folding electronics workbench shown below. Comprising a plywood box with a fold-down front face, it packs into its small dimensions a soldering iron, variable power supply, multimeter, work light and electronic components. Overall size is 420mm (16½in) high, 200mm (just under 8in) deep, and 560mm (22in) wide. A similar approach could be taken with a cabinet housing a small vice, cutting tools and some files – fine for making trim panels and the like.

Amongst my first major work on a car was performed in the open car park of the accommodation in which I then lived. It wasn't very comfortable, and I needed to pack up all my tools and ensure the security of the car each time I

My first real home car workshop, located under an elevated house. The bench carries a small drill press, grinder with belt sanding attachment, a piece of steel channel being used as an anvil, and a 150mm (6in) vice. Power is supplied from a cable hanging down from above.

finished work, but it was possible. Funnily enough, at the time I was a teacher in a smallish town, and my students used to come to the car park on their bicycles, no doubt to see what bits of his car the mad teacher had on the ground that day! If you're enthusiastic enough, you can work almost anywhere.

## A VERY SMALL WORKSHOP

My first real workshop comprised a space only about 3 x 3m (10 x 10ft). However, it wasn't quite as cramped as it sounds because it was positioned under an elevated house, and so was open to the outside air on one side, and under-house car parking locations on two other sides. There was room for two cars to be parked under the house – but it was nothing like a double car workshop, because those cars needed to come and go as required! My workspace was just the bit tucked into the corner.

The workshop area was dominated by my workbench – the same 1.5 x 1m (4½ x 3ft) bench that you'll see used in my following workshop layouts as well. In this case, the bench was pushed against a wall across which were positioned shelves. Mounted on the bench were a small drill press, grinder with belt sanding attachment, a piece of steel channel being used as an anvil, and a 150mm (6in) vice. Power was supplied by a cable hanging down from

A folding electronics workbench – a good approach when you have no workshop space available at all.
(Courtesy Morten Nisker Toppenberg)

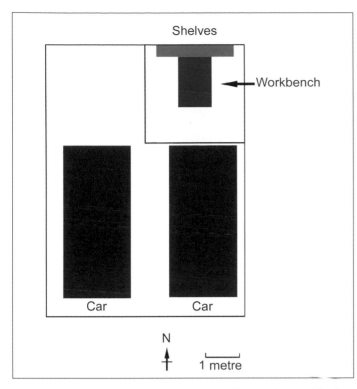

**The layout of that first small workshop. The workspace I had available is just the bit tucked into the corner – the cars needed to come and go as required. Note how the workspace is dominated by the 1.5 x 1m (about 4½ x 3ft) bench.**

above, and a powerful light was located directly above the workbench.

While obviously large items could not be handled in such a small work space, this combination of basic machine tools, a sturdy bench and a good vice allowed a lot of quality work to be performed in this area.

## A SMALL WORKSHOP

A few years later, I found myself in the situation where again I needed to organise a small home workshop. But this time, I needed to fit in a lot more equipment I'd accumulated along the way. I'd moved to a rental house that had available a 7 x 4.2m (about 23 x 14ft) single car garage. It had a large roller door at one end and a small personal access door on a side wall towards the other end. So how do you organise an effective home workshop in a space like this?

The first decision I took was that the new workshop would not be able to house a car – not as a garage and not for being worked on. A standard-sized single car garage is simply too small to be effective as a workshop *and* a place where you can work on a car – and if you try to achieve these dual outcomes, you'll just be very frustrated. An option when you're in this situation is to leave one end of the shed

available for the car, so that you can poke one end of the car through an open door and still be able to work on the engine area without being exposed to the weather. But in the case of this home workshop, I deliberately excluded this option as well. I still wanted to be able to work on my car, but that would need to occur outside only when the weather was fine. (But what if you live in a climate where that's not possible? More on this in a moment.)

The decision to not leave one end of a garage suitable for car work is important, because if there is also a separate personal access door, you can treat the end of the shed with the car access door as just another wall. Depending on the shape of the shed, this can increase the wall area by about 20 per cent. And when you have only a small workshop, you want to use every bit of wall space for storage! (However, if taking this approach, ensure that the major door can still be opened – in my case, if there was to be sufficient ventilation, I needed to do that when welding. Opening the door also added plenty of light.)

The first step in planning is to work out what needs to go where in the workshop. In my case the list of major items comprised:

* workbench
* power tool bench (holding hydraulic press, grinder, drill press, belt sander)
* vertical mill
* lathe
* vertical bandsaw
* oxy acetylene welding bench

Of these, I preferred the workbench and the welding bench to be located as 'islands' – that is, not against a

**A lot packed into a 7 x 4.2m (about 23 x 14ft) single car garage. Inside the space there was enough room for an island workbench, welding bench, power tools bench, and a lathe, mill and vertical bandsaw.**

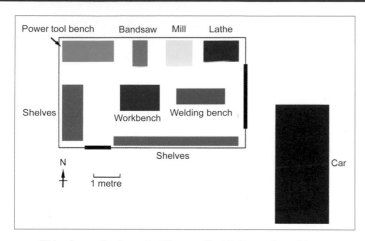

**This shows the layout of the smallest fully-equipped home workshop that I've had. It was housed in a 7 x 4.2m (about 23 x 14ft) single car garage. In this case I decided that all work on the car (purple rectangle) would occur outside. Note the two workbenches are islands, allowing large objects to overhang them.**

wall. Conversely, the bandsaw, mill, lathe and power tools bench could all go against walls. In addition, I had multiple racks of shelving – too many in fact to fit into the shed. (Luckily, I had access to a storage container at another location that I could use for items that wouldn't fit into the small workshop.)

Another issue was that the heaviest items needed to be located nearest to the major door – simply so that when they were lifted into the shed by my hydraulic engine crane, they'd be moved only the shortest possible distance.

After examining the space and looking at my gear, I came up with a plan – shown in the diagram above. Against the north wall, from left to right, were the power tool bench, the bandsaw, the mill and the lathe. The main workbench and the welding workbench were both positioned in the middle of the space, allowing large and/or long items to overlap the edges of the benches without fouling anything. Storage shelves were positioned on the other walls.

Some advantages of this layout included that the equipment that created sparks (eg the belt sander and grinder on the power tool bench) were located well away from the oxy acetylene welding equipment. They were also distant from the precision-ground lathe and mill ways – the latter you want to keep away from thrown abrasive particles. As mentioned, the welding bench was also near the main door, that could be opened for ventilation.

The plan worked well – but what was less welcome was the dearth of storage space. More than anything else, I'd have liked a lot more wall space for shelving! But without that being available, I walked around the shed working out what could come off the floor and be suspended from the roof, mounted high against walls or otherwise be stacked out of the way. Handled in this way were car ramps, jack-

stands, clamps, plastic tube, timber – anything I could get off the floor and up high. Where I could, I used shelves from floor to the roof – these shelves contained mostly tools and fasteners.

Lighting was achieved by hanging HID (high intensity discharge) lamps from the roof. For best spread, these should really have been mounted higher, but they were OK for relatively short-term use. The shed came with just a single power outlet, but I added to it a portable, safety-switched (ELCB) quad power board and then ran from this board cables for the lights and machine tools. The power supply to the island workbench was via an extension cord hanging down from the roof.

The contents of the workshop were certainly packed in tightly but the space was absolutely fine to work in. I had room to move and position work-pieces, and the lighting, power and ventilation were all quite satisfactory. But that was one workshop that I simply had to keep clean and tidy …

As I've said, the workshop described above worked well for me. In fact, in the year we were at that house, I fitted to one of my cars a large intercooler (including fabricating all the plumbing) and big brakes, working just outside the roller door. But that's in sunny Australia – what if you live in a climate where you really must work inside? How do you fit a car in a small space with workshop facilities as well?

The second diagram of the small workshop (below) displays what I'd do in that situation. A workbench is located against the western wall, with a power tool bench located alongside. This 2m (6ft) long tool bench is large enough to carry a small drill press, a grinder or belt sander and another power tool of your choice. A compact welder could also be placed under this bench, with a thick steel sheet laid over your normal workbench when welding. Shelving is confined

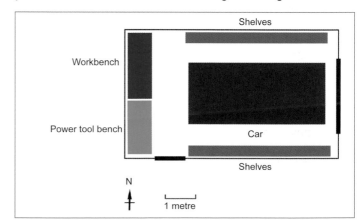

**Here's the same 7 x 4.2m (about 23 x 14ft) space but this time shown with a car positioned inside the workshop. There's still room for a workbench and power tool bench. Space around the car is now sufficiently tight that only very narrow storage shelves can be used each side of the vehicle. If you were working on a motorcycle, there would be room for much deeper side shelves.**

to narrow shelves down each side of the space. As with each of these layout diagrams, I've shown a car that's 4.5 x 2m (about 15 x 6½ft) – a larger car will make things very tight indeed. Note that the car should be protected when using tools like the grinder and welder – it's quite close to your working area.

## A MEDIUM-SIZED WORKSHOP

At one of the houses I've lived in I was fortunate enough to be able to build a medium-sized (some might call it large) home workshop from scratch. It comprised at 14 x 6 x 3m (about 46 x 20 x 10ft) steel shed with a concrete floor. Two car access doors were positioned at one end of the space, and with this design I chose not to have a personal access door. This was the first workshop I built from scratch, and I spent a lot of time planning its internal layout.

With a workshop that is long and skinny, and has doors at one end, some layout decisions simply fall into place. For example, I wanted room in the workshop to have two cars – one being stored and one being worked on. The space for the two cars was at the end of the workshop with the two main doors. That way, each car could enter or leave the workshop separately. (However, so that there remained room for the two cars to be located next to one another, intrusion at this end of the workshop into floor space by wall shelves and cupboards needed to be kept very low.) By default, the 'workshop' part of the shed then became the other end of the shed.

As with the small workshop layout described earlier, the next step was to think-through those workshop items that required the smallest space behind them, and so could be pushed up against walls. For example, my island workbench should not be against a wall. On the other hand, my lathe required no rear access at all. On this basis, I placed against the western wall the lathe, mill and bandsaw. I also have small bench I constructed – it's wide but shallow in depth. I had previously mounted on it a little anvil, bench grinder, drill press and hydraulic press. By replacing the anvil with a bench sander, I could add another collection of tools for which rear access was not required. Therefore, mounted

in a line from left to right could be: bench with hydraulic press, bench grinder, drill press, belt sander; and then on their own stands: bandsaw, mill and lathe. I made some measurements and found that this collection nicely fitted across the western wall of the workshop.

The next decision was where to put my sheet metal bender. I had bought this sight-unseen, and it proved to be far bigger than the picture in the ad had suggested. That made it an even better bargain, but it also meant it took up a lot of space and was very hard to move. It was also very heavy – perhaps three-quarters of a ton – and so I decided that this should be placed in the workshop on the part of the concrete slab that was not built on 'fill,' but instead was on more solid ground. By default, that put it against the southern wall. Note that space also needed to be left for the moving leaf to open, and a counter-weight to descend behind the folder.

The next two decisions were where to put the two main working benches. As with the small workshop layout described earlier, these comprised an island workbench and a welding (oxy acetylene) bench. After chalking their shapes on the floor and walking around them many times, I decided to place them parallel to one another, near to the rear wall of power tools. This gave good access from the welding bench to the workbench with the big vice, and also kept the welding flame well away from where paint and inflammables were stored. The location of the oxy bench was also distant from the spark-producing grinders. Finally, lots of storage shelves were added to the remaining free walls.

Note how in this layout the island workbench is centrally located with regard to the main tools. For example, when fabricating a bracket, it was easy to cut the steel stock to size

**This is a at 14 x 6m (about 46 x 20ft) home workshop. Two cars can be parked in the space closest to the camera. This photo is taken at night – note the evenness of the lighting.**

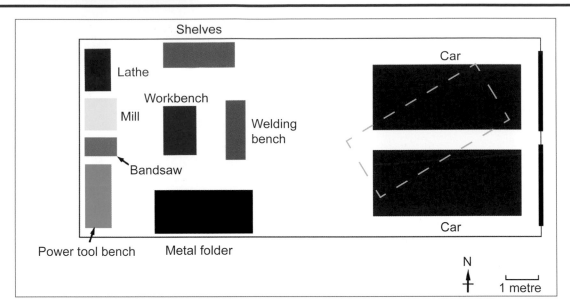

This is a plan of the 14 x 6m (about 46 x 20ft) home workshop. It has room for two cars and a good range of workshop tools down one end. The central location of the island workbench within the tools gives quick and easy access to bench-mounted power tools, vertical bandsaw, lathe, mill and metal folder. A single car could also be parked diagonally (dashed line) or at right angles across the space, leaving plenty of room around it.

in the vice on the workbench, take two steps and de-burr it with the belt sander, return to the workbench to mark it out and centre punch it, then take a few steps to the drill press to drill the holes. In fact, when making things, I never had to take more than four steps to reach the next machine. Thinking things through in this way is very important if you want to make the workshop as productive as possible. It also helps in planning the layout of fatigue-reduction rubber mats on the floor – typically, the required area of mats is reduced.

This workshop worked very well, and in fact I made no changes to the layout over the next five years. In use, I often removed the second (stored) car and then parked the car I was working on further forward, nosing it in diagonally. This gave plenty of room around whole car. In fact, the only shortcoming that arose was a lack of ventilation when welding – the two large roller doors being at the other end of the space, and no opening windows being provided.

## A LARGE WORKSHOP

When the time came to move again, I rather reluctantly left behind the workshop described above. However, at least I could take all the contents with me but for the huge sheet metal folder, that I sadly sold. (I've since bought a much smaller folder.)

I was again able to construct a home workshop building from scratch, but this time I decided to go in a different direction – I went high! Again, this building comprised a steel shed and a concrete floor, but rather than having the 3m (10ft) wall height of the previous design, I went for a 5m (about 16ft) high wall, with a floor area of 12 x 8m (about 39 x 26ft).

The major rationale for decreasing the floor area over the previous design, but increasing wall height, was that lots of the cost of a workshop of this type is the concrete

slab floor. This means that the cost of the workshop goes up rapidly with increasing floor area, but much less rapidly if the building increases in height. So despite the building being large in volume, the cost was less than building a larger area, smaller height shed. (Would I do the same again? Perhaps not – more on this in a moment.) Note that this workshop was located in a rural area, and so the tall building height was able to be accommodated within planning laws. In an urban area, you're much less likely to be able to build such a tall structure.

To arrive at the 12 x 8m (about 39 x 26ft) floor area took a number of planning steps. I initially started making some sketches, working from the previous 14 x 6m (about 46 x 20ft) design. However, the location of the shed meant that rather than having car access doors at one end of a long skinny design, these doors would need to go in the side wall. But with these doors being in a side wall, a 6m (20ft) width meant that any cars in the shed would be parked on this short axis – and that was a problem. Why? Well, a length of 6m (20ft), while enough for car storage, doesn't provide sufficient space ahead of (and behind) the car to allow easy working on it.

So then I went up in width to 7m (23ft), then 8m (26ft). An 8m (26ft) width, I figured, would give sufficient room either end of the car, even with a workbench placed along one of the long shed walls. However, a shed of this size has an area of 112m$^2$ (1200ft$^2$) – and the cost of a concrete floor was starting to get rather high. The length of the shed was then dropped back to 12m (39ft), giving a 12 x 8m (39 x 26ft) footprint – an area of 96m$^2$ (about 1000 ft$^2$).

The next step was to consider doors. As I said, a deficiency of the previous workshop was that ventilation was insufficient. That was primarily because the sole doors were at one end of the long, relatively narrow workshop. In the new design, the main doors were to go in the long wall, so the 'doors at one end' effect would be less of an issue. But how many doors should be fitted?

I settled on two large car access doors in the side wall, positioned side-by-side. The second door was primarily to provide ventilation – in fact, I initially set up my welding bench in front of it, and so by opening the door, I could potentially have a lot of air flow. Finally, despite my being quite happy not having a personal access door in the previous workshop design, my wife requested that we have one in this shed – and I am glad she did.

Now, how to arrange items within the space? This workshop – which is my current one – has undergone several variations in interior layout, so I'll describe only the current configuration. Over the years it's become rather full!

Unlike the previously described workshops, the first step was to organise storage. This occurred first because I chose to use pallet racking from floor to roof across the whole of the western wall, and in smaller lengths on the north and south walls. The pallet racking, that stands some 4.9m (16ft) tall, provides an enormous amount of storage. By suitably specifying the height of the lowest crossbeams, it's possible to place this storage well above working areas, so detracting only a small amount from floor space (just the footprint of the legs). Wooden slats were screwed to the pallet framing to provide a flat storage area in each bay. To permit the large uprights to be easily put into place, the pallet racking was installed before any other items were located in the workshop.

As with the previous two workshops that I've described, the lathe, mill and original power tool benches could all be positioned against walls. The island work and welding benches were positioned with free area round them – and in the case of the welding bench, with spot, MIG, and oxy-acetylene welders nearby.

Nearby is a second power tool bench. This bench carries the horizontal bandsaw that is used to cut long stock to length. It's positioned so that the stock can extend out through the open door. Only two steps away from the island workbench is my main tool chest, and one step away

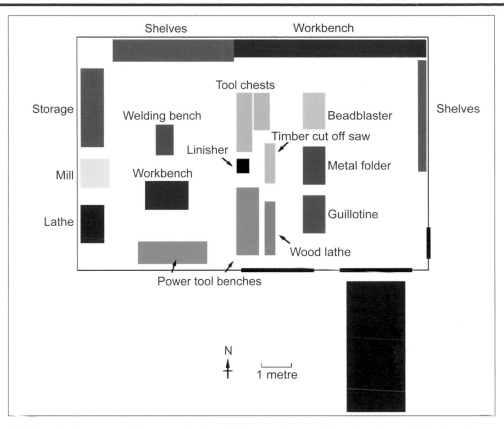

The layout of my current home workshop. This uses two workbenches (one an island and one against a wall), two power tool workbenches, and a wide array of tools and machines. Room for the car in the workshop is being squeezed by equipment acquisitions, but is still adequate.

is my large Radius Master linisher. The area I have so far described is my main working area.

There's also a second 'avenue' of tools to the east – a second tool chest, a woodworking lathe and a timber-cutting drop saw. On the other side of the 'avenue' is a bead-blasting cabinet, a metal folder and sheet metal guillotine. Across the northern wall is a long timber bench. The last three items I've bought – the bead-blasting cabinet, metal folder and sheet metal guillotine – have really eaten into the space I'd set aside for the car on which I am working … too many tools!

Would I go as high in workshop design again? Probably not. The tall pallet racking can hold a great amount, but accessing it is via a ladder. This means in turn that the highest storage is suitable only for items that are light enough to be carried up a ladder. Even small crates become quite heavy when loaded with mechanical items, and so the highest shelves in my workshop have turned out to be packed with a lot of small crates and very light items. Having said that, part of the original thought behind a high workshop was that a mezzanine floor could later be placed at one end of the space. (In fact, when it was being laid, I had the concrete slab strengthened where the poles that supported this mezzanine were to go.) However, over time I moved away from the idea of a mezzanine, primarily because it would lower roof clearance over my work area and reduce natural lighting.

Looking at the layout now, I think I need to get rid of the long workbench against the northern wall and move some of the major fixed tools (eg the metal folder, guillotine and bead blaster to this wall, so freeing up some space again for a car to fit more easily into the working space.

## JESSE HOLMAN'S WORKSHOP

Jesse Holman is in the throes of equipping a very large home workshop. The workshop comprises two adjoining spaces, one 17 x 4m (56 x 13ft), and the other a substantial 17 x 11m (56 x 36ft). Ceiling height is 4m (13ft). The smaller space is for storage (and also houses the air compressor) while the larger area is the main car workshop and is insulated and heated. An L-shaped workbench is positioned in one corner and comprises recycled kitchen benches and laminated bench tops. The upper cupboards from the kitchen are also mounted here, giving plenty of storage for lightweight items. Jesse will probably find that he needs another, heavy-duty workbench on which to mount a vice – but there's plenty of room in the workshop for that.

On one side of the benches is positioned a mobile storage rack and then a hydraulic press and mobile engine

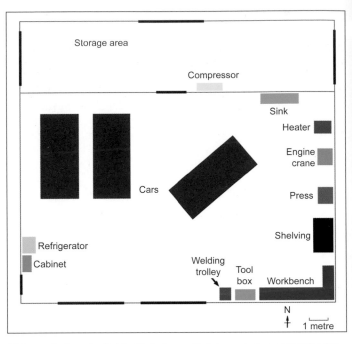

This workshop is divided into two adjoining spaces, one 17 x 4m (56 x 13ft) and the other 17 x 11m (56 x 36ft). The smaller space is for storage (and also houses the air compressor) while the larger area is the main car workshop and is insulated and heated. Ceiling fans are also fitted.

crane. On the other side of the bench is a mobile toolbox and the welding trolley. Compressed air is piped to outlets within the workshop (including a wall-mounted retractable air hose), suspended LED lighting is used and the floor is concrete. There is massive room for more equipment in this work space – Jesse plans to install a two-post hoist, drill press and mobile welding table/workbench. At the other end of the workshop there's room for two cars, and Jesse has also located a fridge and cupboard near the personal access door.

## RUSSELL FERGUSON'S WORKSHOP

Russell Ferguson packs a huge amount into a limited space. This is a well-equipped workshop that has been built up over many years. The workshop is 9 x 6m (about 30 x 20ft) with a single car access door located at one end. The workshop features a compact bench-mounted lathe, a

Jesse Holman's home workshop has plenty of space left for the installation of more equipment. This section of the workshop is a substantial 11m (36ft) wide, uses suspended LED lights and a concrete floor. Note the recycled corner kitchen cabinets and benches to give lots of storage. A second window is yet to be installed.
(Courtesy Mallory Lauren Photography)

Left: This wide-angle view shows Russell Ferguson's workshop. The space is 9 x 6m (about 30 x 20ft). All workbenches and machine tools are positioned along walls.

Behind the wood burner and two comfortable chairs are a bench with vice, small lathe, mill, drill press, compressor with vertical tank and storage chest. Note the fluorescent lighting over the bench.

How to use a wall! From left can be seen a storage chest, trolley jack, jack-stands, hydraulic press, hung-up mains power extension cords, a grinder, engine crane, welding trolley and welding bench. Note the road signs used as decoration.

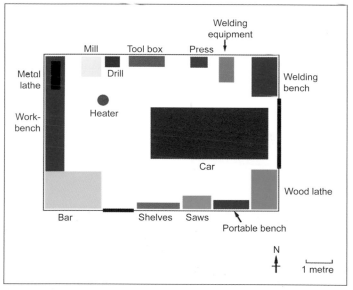

Russell's workshop packs a lot into its dimensions. Note how in addition to space for working on the car, the workshop also features a compact lathe, a mill, drill press, hydraulic press, welding equipment and dedicated welding bench, wood lathe and powered wood saws, a workbench – and a bar!

Behind the parked car can be seen a wood lathe and wood-turning chisels, a table saw and a bandsaw – and lots of stored ropes, oil and equipment. Note how the bench in front of the wood lathe can be moved when access to the lathe is needed.
(All photos courtesy Wayne Suffield)

Bill Sherwood's well equipped workshop. Top left: a large lathe (large for a home workshop, anyway), extensive tool chests and, in the foreground, a well-braced steel workbench. Notice also the web-access PC to the right – useful for checking on a technique or updating a discussion forum about your project. Centre left: The rack on the wall is for holding steel bar and tube stock. Always plan a workshop to have plenty of space for the storage of materials. In the foreground is the V8 race car that Bill built in this workshop, which commemorates his father's racing Cortina. Bottom left: A bay is set aside for a two-post hoist. Behind this are supplies of oil and paint, another tool chest, a large hydraulic press and a small grinder. Top right: A spit allows a car body to be rotated for easy work on any fixed panel. Note also in the foreground the indexing heads and lathe chuck. (All photos courtesy Bill Sherwood)

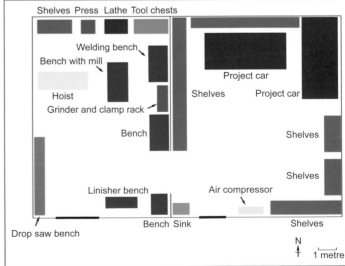

The total area of this workshop is 17 x 11m (56 x 36ft). It houses a multitude of benches, including two island and two wall benches, plus a mobile steel bench. To the left is a two-post hoist and the hand and machine tools. The other side is largely devoted to storage – two project cars (both small vehicles), plenty of shelving, along with the air compressor and sink.

mill, drill press, hydraulic press, welding equipment and dedicated welding bench, wood lathe and powered wood saws, a long workbench – and a bar!

## BILL SHERWOOD'S WORKSHOP

Bill Sherwood's workshop is in two parts – the original construction and then an added construction that doubled the area. The total area is 17 x 11m (56 x 36ft). In the western side is located a multitude of benches, including two island benches and two placed against walls. One steel bench is mobile within the workshop. Bill uses a two-post hoist and also places in this side of the workshop the hand and machine tools. The eastern side of the workshop is largely devoted to storage – two project cars (both small vehicles), and plenty of shelving for parts. The air compressor and sink are also located in this side of the workshop space.

## RULES OF WORKSHOP ORGANISATION

So where in your home workshop should you locate things?

| This... | Not near this... | Because ... |
|---|---|---|
| Grinder Bench belt sander Friction blade cut-off saw | Lathe Mill | Abrasive articles can wear the accurately ground machine tool surfaces |
| Main workbench Bench belt sander Oxy acetylene workbench Arc welding gear Friction blade cut-off saw | Paint Stored fuel Aerosols | Danger of fire |
| Fluorescent and AC LED lights | Drill press Bench belt sander Mill Lathe | At certain speeds, the rotating tools will appear stationary |
| Air compressor | Working areas | Noise |

| This ... | Near this ... | Because ... |
|---|---|---|
| Workbench | Most commonly used hand tools (e.g. screwdrivers, spanners, sockets hammers) | Frequent access required |
| Workbench | Drill press Bench belt sander | Frequent access required |
| Oxy acetylene gear | Workbench vice Anvil | Heating and bending requirements needs good access |

| Machine | Space needed | Because ... |
|---|---|---|
| Workbench | If an 'island' design, space is needed around all sides. If an 'along the wall' design, space is needed primarily in front, and to a lesser extent, at ends. | Large, flat items will often overhang the bench |
| Friction cut off saw, horizontal bandsaw | Little space needed in front, but very long clearance needed at one end. If mitre cuts are to be made and the stock will need to swivel, long clearances needed at appropriate angles (e.g. 45 degrees). | Chopping the end of long lengths of material |
| Vice Drill press Mill Lathe Hydraulic press | Little clearance usually needed in front (enough to stand comfortably). End clearance at working level much longer. Any smaller machines that mount in the vice (sheet metal rollers, tube benders) will need clearance for stock and actuating handles. | Unless they are long and skinny, most items will be relatively small and kept within the confines of the working area of the machines. |
| Bench belt sander | Fairly tall vertical clearance Reasonable clearance width | Most items will be small and easily handheld. Sometimes, when grinding 'along the grain,' taller clearance will be needed. |
| Sheet metal folder Sheet metal rollers Sheet metal guillotine Pipe bender | Depends on most common application and size of machines | Clearance for handling sheets and moving handles. |

**Chapter 11**

# Designing a home workshop

In this chapter, I want to look at the universal aspects you should aim for if designing a home workshop from scratch. I won't be spending a lot of time looking at the actual construction process, primarily because that will vary, depending on the local climate and building methodologies. For example, where I live, by far the cheapest and most effective approach to building a workshop is to use a galvanised sheet steel shed with a concrete floor. Such a structure is durable and flexible. However, it's also uninsulated and so hard to heat and cool. In a mild climate, that aspect doesn't matter much, but in a climate of extremes, it may make the workshop unusable for half the year. So leaving aside the actual construction methodology, what should you aim for in your workshop design?

## SHAPE

While buildings can have any plan shape that's desired, most home workshops will be rectangular or square in form.

A long, narrow rectangle is effective if the access door for the vehicle is at one end of the space. In this approach, the space for working on the car is at one end of the rectangle, and the workshop tools and benches are at the other end. Especially if a lot of your work is under the bonnet (hood), this works well. However, if doing bodywork, this approach isn't so good as clearance each side of the car is quite limited.

If the rectangle is not so narrow, the car can enter via doors positioned on the long side. In this approach, the width of the workshop must be sufficient to provide enough work room at the front (or rear) of the car. At minimum, you need a workshop width that is 2m (6ft) wider than the longest vehicle you intend working on.

A square plan allows greater versatility in layout. You can choose to have plenty of space in front and behind the vehicle, or even have the vehicle positioned in the middle of the workshop, with the tools and benches located around the walls. For example, if you're intending to use wall benches (as opposed to an island bench), positioning the car well away from the walls will be required and so a square plan works well.

When looking at different shapes of floor plan, remember that the car needs to be easily able to enter and leave the workshop. It's easy to end up with a design where the car has plenty of room once it's inside the space, but entering and exiting is little awkward. If you are making changes to the vehicle (think suspension or engine tuning changes), you want to be able to quickly road test the results before re-entering the workshop to make more changes.

In many cases, the shape of the workshop will be dictated by the available area of land. However, even if you're working within that constraint, you often have flexibility in the location of the doors – and these will be a major influence on how the workshop can then be internally organised.

## HEIGHT

I have worked in home workshops that have internal heights that vary from 2m to 5m (about 6ft to 16ft). Higher workshops have these advantages:
* more clearance for swinging objects (eg a long length of steel tube)
* the ability to have artificial and natural light sources positioned high up, so resulting in a more even spread of illumination
* cooler in hot climates
* much more storage space if high racks are used
* plenty of clearance for a hoist
* may be able to have a mezzanine floor installed

However, higher workshops also have these disadvantages:
* increased build cost
* very high workshops may not be able to be built due to local regulations
* difficult to heat
* high-mounted storage may be hard to access, particularly for heavy items

I don't think that there's just one simple answer to the question of best workshop height, so if you are building from scratch, consider this aspect very carefully.

## LIGHTING

Lighting is the one of the most important aspects to get right in a home workshop. So many home workshops are dull in daytime – and duller at night!

So how much light is needed? This is a very important question, and one that can be easily answered. Furthermore, with light meters now cheaply available, you can also make measurements to see how effective the lighting actually is. Illuminance is measured in lux. There are recommended values of maintained illuminance for various activities, with

**Jesse Holman's workshop has more than enough space for a car to enter and then be parked at an angle. Note also the bright and even illumination provided by the suspended LED lighting. (Courtesy Mallory Lauren Photography)**

A tall workshop allows the use of commensurately tall racks for storage. However, unless you have a forklift, accessing the top shelves is difficult.

the table below showing some International Commission on Illumination (CIE) suggestions.

| Location | Illuminance (lux) |
| --- | --- |
| Hospital ward at night | 1 |
| Cinema auditorium | 50 |
| Toilets | 100 |
| Garment manufacture – sewing | 750 |
| School classrooms | 500 |
| Kitchen work areas | 500 |
| Supermarket | 750 |
| Instrument assembly | 1500 |
| Operating theatre (local lighting) | 100,000 |

Looking at the table, you can see that you need an achieved illuminance of at least 500 lux – *measured at bench level*. (Note that normal handheld light meters measure in lux.) In addition, you'll need localised higher illuminance at machine tools (eg a lathe) and in places that are otherwise shaded from the overall lighting (eg under a car).

Natural lighting can be provided by windows, large doors that are opened when the workshop is in use (if the climate permits this) or skylights. If the workshop comprises a shed of some type, skylights are easily and cheaply provided by using translucent sheets. These panels are the same size and shape as the normal roofing sheets, so their installation is straightforward. In most weather conditions, these panels provide a very large amount of light. However, if you live in a hot climate, these panels will also admit heat. One approach is to use framed panels of shade cloth over (or under) the skylights in summer. Providing natural light through windows has the advantage that they can be opened for increased ventilation, but the disadvantage that they take up wall space that could otherwise be used for storage. When considering the orientation of the workshop, remember that skylights work best on south-facing roof panels in the northern hemisphere, or north-facing roof panels in the southern hemisphere.

In my mild climate, I use skylights for main daytime illumination, with doors opened as required to provide further light and increased ventilation.

In addition to being bright, when selecting artificial lighting you also want:
- lights with a high luminous efficacy (ie lots of lumens output per electric watt input)
- good colour rendering (so there's no colour cast)
- mounted in a luminaire that provides high luminous intensity (ie directs the light appropriately)

At the time of writing, commercial LED lighting best meets these criteria. If you have a high roof, use those luminaires designed for commercial workshops and warehouses (these are sometimes called 'high-bay' lights). If you have a lower roof, use strip-type LED luminaires. It is better to use a larger number of widely distributed lights than just a few lights. This is because a small number of bright lights will give darker shadows and so make seeing things more difficult. Localised intense lighting (such as at machine tools) should use focused, narrow beam light sources.

When designing the lighting system, don't forget to provide adequate external lighting around the workshop. This may be needed for security purposes, or in mild climates, when work spills outside the workshop.

For one of my home workshops, built before LED lighting was widely available, I did lots of experimentation with compact and linear fluorescent tubes, and metal halide and sodium vapour light sources. (Note that most metal halide and sodium vapour lights need their own specialised power supplies, usually built into the luminaire.) Measured lux values at bench level varied from 72 to 650 lux, depending on the lighting source and the reflector being used. In the end, I went for 12 150W metal halide lights mounted in large reflectors. I bought the bulbs new, but managed to source the control gear and reflectors secondhand at low cost. I operated the 12 lights in two banks of six. This gave a near-even 650 lux right through the 14 x 6 x 3m (about 46 x 20 x 10ft) workshop. The lighting was so good that, after switching the lights on in the late afternoon, I often didn't realise that

Constructed before the widespread availability of LED lighting, this 14 x 6 x 3m (about 46 x 20 x 10ft) workshop used 12 150W metal halide lights, operated in two banks of six. This gave 650 lux of illumination (measured at bench level) right through the workshop.

Here lights in front of the car access door have been suspended from supports that project out from each wall. Taking this approach still allows the door the required clearance to slide on its tracks against the roof. (Courtesy Jack Olsen)

night had fallen, and walked out in surprise to the utter darkness.

## VENTILATION

Ventilation in a home workshop is very important. Ventilation can comprise natural ventilation or forced air ventilation – or both.

Natural ventilation can occur through the opening of doors or windows as required. It can also utilise spinning, wind-driven extraction vents located on the roof. Natural ventilation will occur only if there are both air inlets and outlets, and a

pressure difference between the two is created. For example, spinning ventilators will draw air out of the workshop, creating a lower pressure that is then filled by air entering the workshop. But to provide *you* with fresh air, that intake airflow must pass you! As an example of how *not* to do it, imagine a workshop with a single window and a door, with both located in a corner. These openings may flow plenty of air, but most of the workshop will still remain quite stuffy. To avoid this sort of effect, intakes for ventilation air must be widely distributed – for example, using multiple grilles or vents in the lower walls.

This was brought home to me in my second custom home workshop, a 14 x 6m (about 46 x 20ft) space. There were two large doors at one end, and my welding bench was located at the other end. On the roof were two wind-driven spinning ventilators. However, even with the doors fully open, the area around the welding bench became quite polluted. What was needed were either ventilation inlets in the walls either side of the bench, or a purpose-built fume extraction system.

Especially in a workshop that you are heating, ventilation needs to be controllable. The easiest way of achieving this is to use a forced-air exhaust fan system that is switched on as needed. The intake may be provided by fixed ventilators or ones that are opened as required.

Using an old woodworking dust collector, I built a powerful exhaust fan system for my welding bench. The two original

I salvaged this heavy-duty exhaust fan from an old woodworking dust collector. The top pipe exits to the outside air and the intake's convoluted tube connects to a long, pivoting arm that allows the mouth of the tube to be positioned appropriately – for example, on the welding bench.

large filter/collection bags were discarded and the fan was mounted on a frame bolted to the workshop wall. The fan exhausted through a grille in this wall to the outside air. A large diameter convoluted tube led from the fan's intake to a long, pivoting arm, mounted high in the air. The pivot allowed the tube to be swivelled over the welding bench when welding was occurring, and tucked out of the way when it was not needed. The hose was made sufficiently long that the intake end could be placed on the welding bench itself. This system worked very well.

I'll have more to say on this topic in the next chapter, but two of the greatest dangers in a home workshop are the toxic gases from engine exhausts, and the gases released when galvanised steel is welded. In both cases you need substantial ventilation, preferably of the forced-air type.

## POWER OUTLETS (SOCKETS)

It's one of those things that sounds pretty simple – where in a workshop to put the mains (line) power outlets. However, if you want the best outcome, it's anything but simple. For example, many people distribute the power sockets evenly around the periphery. But that assumes that the requirement for power is also spread evenly throughout the workshop – and that's not normally the case. In fact, before you can know where to position the power outlets, you need to have a good idea of how the workshop will be organised.

For one of my home workshops, my thinking went like this. (Note: this thinking was done before the workshop was even started!)

*Across the back wall of the shed are positioned the lathe, mill, bandsaw, and a bench housing a sander, grinder, drill press and hydraulic press. Tucked below the bench is an air compressor. Therefore, just across this shortest wall, there are seven machines requiring power. But both the mill and the lathe may well need a second power outlet for each machine. For example, each might need a suds pump to circulate coolant over the item being machined. Each might also need a dedicated work light. So it makes sense to provide both the lathe and the mill with double power outlets. The bandsaw, grinder, sander, drill press and compressor need one power outlet each – so that's two-and-half double power outlets. So, leaving one spare outlet, that adds another three doubles. Therefore, five double power points – 10 outlets – on just this one wall.*

A similar calculation can be carried out for the other walls in the workshop, not forgetting that in the area where the vehicle will be worked on, there should be multiple power outlets towards both the front and rear of the vehicle. To allow power to be accessible away from the walls, I prefer to use power outlets hung from the roof. By plugging into a socket located above your head, cords no longer need to run across the floor or workbenches. In the 14 x 6 x 3m (about 46 x 20 x 10ft) workshop, the final power outlet tally comprised

Suspended power sockets reduce the need for cables to be run across floors and workbenches. However, unlike the one shown here, I normally position them just within reach above head height. (Courtesy Clipsal)

10 doubles, two hanging outlets and a single weatherproof socket mounted outside near the doors.

All power outlets in a home workshop should be protected by appropriate circuit breakers that also incorporate earth leakage circuit breakers (safety switches or ELCBs).

Finally on power, I prefer to have a master switch for power near to a door, and to switch off all power when I leave the workshop. Sometimes this is not what you want to do (eg if you're charging a battery), but I find that on most occasions, when I am not in the workshop, I want all power off. To remind me to do this, I've added a bright neon indicator to the switchboard (it's located near the personal access door) that is illuminated whenever power is switched on.

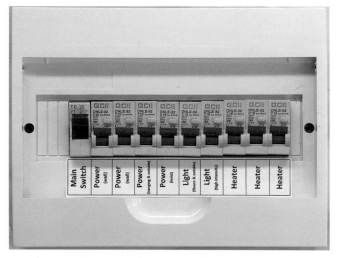

The well-labelled power board in my workshop. I have the board positioned close to the personal access door, and I usually flick off the main switch (red) as I exit.

## SECURITY

A home workshop – especially one that is well-equipped – has a lot of items that are both valuable and portable. Think of all your hand and power tools, for a start. It's therefore wise to carefully consider both security and insurance.

On insurance first, many people underestimate the replacement cost of the tools and equipment that they've built up over a long period. Discuss your home workshop with your insurance company, for example to see if it is adequately covered under an existing home and contents insurance policy. In my experience, many insurance companies are surprised by the value of the contents of a home workshop. Some insurance companies may ask for lists of individual tools and equipment, or require that you mark them with identification.

Security should at minimum comprise adequate locks on all doors (and don't forget that some car access doors are very easy to break into), appropriate warning stickers, and video cameras that record to a memory source. Ensure that video cameras have sufficient resolution to read vehicle number plates and provide facial identification, and use at least two cameras – one inside and one outside the workshop. A simple step is to use movement actuated lights around the outside of the workshop.

## FLOOR FINISH

The choice of surface finish that you use on the floor of the workshop is a topic around which there is a lot of debate.

I prefer bare concrete – but it needs to be concrete that is well finished with a hard and smooth surface. Concrete that has a rough finish will be difficult to sweep, and concrete that has a friable upper surface (it easily crumbles) is an endless source of dust and grit. A major benefit of using just bare concrete is that it doesn't cost anything extra. It also doesn't

The floor finish within Russell Ferguson's workshop varies to suit the use being made of that space. Here a white/yellow painted pattern has been used under the car (the lighter colours making it easier to find dropped parts), with grey painted concrete elsewhere. (Courtesy Wayne Suffield)

require any maintenance, and it's a surface strong enough to take anything that you can throw at it.

And just a word on concrete curing and finishing. The surface strength of concrete is greatly affected by the curing time – in practical terms, the longer the curing time, the better. Curing of concrete is slowed by keeping the surface wet (eg by ponding or spraying). Inadequate or insufficient curing is one of main causes of weak, powdery surfaces with low abrasion resistance. Proper curing also reduces the permeability of the concrete. Aim to keep concrete wet for at least a week after pouring, and ask your concreter to finish the concrete with a smooth surface. On request, some concreters will also 'burnish' the surface, a process where much more mechanical towelling occurs than normal. This further closes up the concrete pores, making the surface more impermeable.

Many people argue in favour of a painted concrete floor. Painting the surface allows spills to be easily wiped up without leaving stains. It also looks better than bare concrete, and if a light colour paint is used, the floor will reflect light, brightening the space. However, painting the floor can be expensive – for a really good epoxy paint, very expensive – and no matter how good the paint is, it will need to be redone every few years. That's especially the case if a car sits on it for any length of time – the tyres seem to develop a magnetic attraction for the paint and will peel it off when you move the car. Keep in mind that coatings fade and/or yellow, so repairs to just small sections will be visible. Concrete can be polished and then clear-sealed, but the sealant is just a transparent paint and so the same maintenance issues occur with this approach. Compared with bare concrete, any painted surface will be more fragile and so more easily scratched or marked.

Interlocking PVC tiles are also available that are rated for garage and workshop use. However, in my home workshop, they would simply not be viable. I often place steelwork on the floor to weld, and I am not sure what red-hot metal would do to PVC floor tiles – but I can't imagine it would be good! Liquid spills also penetrate along the tile join lines. Finally, PVC floor tiles are expensive.

I have also seen ceramic floor tiles used in home car workshops, but despite the practicality of ceramic tiles in home use, I cannot see the same applying in a workshop. After all, drop a large hammer on the floor even once and you'd have a cracked tile. If you're on a finite budget, the expense of ceramic tiles is also large.

However, note that it is possible to mix and match the above approaches. You could use bare concrete for the area with the major fixed tools, a painted floor where you will be working on the car, and PVC interlocking tiles in a rest and relaxation corner.

## MOVING MACHINE TOOLS

Good quality machine tools are large and heavy. In fact, if you

If a home workshop is being built from scratch, it pays dividends to keep the concrete wet for at least a week after it has been laid. Doing this will result in a surface that is harder and less permeable.

Really heavy machine tools, like this large folder, are best moved into the workshop with the help of a forklift or skid steer loader.

are sourcing machine tools secondhand, the best bargains to be had are often the largest and heaviest! But how do you get such tools into your workshop?

For machine tools that are relatively compact and weigh less than about 250kg (550lb), use a hydraulic engine crane. These cranes are sufficiently cheap that one should be in your workshop anyway, and they're more than capable of lifting a lathe or mill off a vehicle and transporting it to your workshop. As always when moving heavy things, take it carefully, and always double-up on the lifting straps you are using.

For very heavy items like large lathes, mills, metal folders and guillotines, the approach that I have taken is to hire a rubber-tracked, skid-steer loader (often called a Bobcat). The ease and precision with which a good Bobcat driver can place tools in a workshop is amazing, and most medium-sized Bobcats are good for lifting a ton.

## MACHINE SET-UP

Machines with accurately ground ways, like lathes and vertical mills, should be carefully set up, primarily so that there is no twisting of these ways. Apart from machining test pieces and then measuring, the only way of achieving this is to use an engineering quality bubble level. These levels are much like normal levels except they are extraordinarily sensitive – for example, the bubble will move across the full viewing scale when just a normal sheet of paper is placed under one end! Bought new, these levels are very expensive, but they do appear secondhand at much lower prices.

The aim with the level is not necessarily to get the machine level! Confused? Don't be. Imagine the level placed across the machined ways of a lathe, at right-angles to the lathe's longitudinal axis. The bubble of the level indicates perhaps that the lathe is not absolutely level in this direction. Note this bubble reading, and then make the same

measurement at the opposite end of the lathe bed. If the bubble shows a different level, the bed of the lathe is twisted. It's for dialling-out twist of this sort that the level is used.

The first step is to use packing to ensure the machine tool can't be rocked or wobbled by hand. Then assess the twist in the lathe bed (or in the milling table) and counteract it by placing further packing under the points of the machine support that are low. It's rather disconcerting to see how placing a very thin shim under (say) one leg of the stand of a lathe can result in a clear change in the twist of the bed! What looks immensely rigid and strong actually isn't.

If the machine tool is not bolted to a stand, place the shims directly under the machine. If the machine is bolted to a stand and shimming the stand does not correct the situation, consider placing the shims between the stand and the machine. I use a tyre lever and a block of wood as the fulcrum to lift the corner under which the shim is to be placed. Shims can comprise steel or aluminium – if using the latter, make sure that the shim supports a large area. For the very thinnest shims, use brass shim stock.

A lathe being set up with the use of an engineering bubble level. The arrow points to a sheet of thin paper being used to assess the degree of bed twist.

**Chapter 12**

# Safety

Home car workshops kill or maim many people each year – that's a sad fact. So while workshop safety isn't a topic on everyone's lips, it should be.

## A FRAMEWORK FOR ASSESSING SAFETY

A few years ago, I went to an Occupational Health and Safety training day. Yes, I know what you are thinking: a course designed for the lowest common denominator. "Don't smoke when you're filling your car with fuel" – stuff like that. But I must tell you, I actually found the course extremely thought-provoking. In fact, I'd go further than that – I think my risk of injury or death when I am working in my home workshop has clearly been lessened. Let me tell you why – and there are really only two ideas that underpin that increase in safety.

## CALCULATING RISK

Early in the day the course trainer showed us a table (see below). Down the left hand vertical axis was an evaluation of how severely someone could be injured if an accident occurred. Across the top horizontal axis was the likelihood of that accident actually occurring. Within the chart was a rating from 1-6, with the lower the number, the greater the risk.

Now what I found very interesting was that the trainer asked us to assess a particular risk – that is, to pick the correct number in the table. The risk he asked us to assess was that of coming in contact with the electricity running through the power cords he had connecting the wall power socket to his laptop and projector. Hmm, OK – I looked at the table. On the scale of 'likelihood,' I figured that the chance of anyone getting hold of an exposed conductor was 'very unlikely.' However, if someone did do that, I figured it would be pretty bad – so I picked as the outcome 'kill or cause permanent disability or ill health.' The table risk number? I got '3.'

But this is where it gets interesting – because I was wrong. Sure, the likelihood of anyone grabbing a conductor was 'very unlikely' – but if they did, the chances are that they would not die. Why? Because the trainer had a safety switch (an earth leakage circuit breaker – ELCB) plugged in at the wall! In fact, to go further, he pointed out that if in fact the risk number actually was '3,' none of us should have been in the room! Instead, an acceptable risk was indicated by 4, 5 or 6. His chosen number was '6' – low risk indeed. And of course, he was right.

Think of the hazardous tasks you perform in a home workshop – welding with oxy-acetylene gear, for example. I am experienced in this area and I wear the correct gear, so I'd rate my chances of getting burnt as 'Unlikely – could happen but very rarely,' and the outcome if I did get burnt as 'First aid needed.' The resulting risk number is '5' – and that's fine.

But take a different job, like using a metal-cutting vertical bandsaw. Now I'd rate the chance of injury higher than a burn from the oxy – with a bandsaw, it's just so easy to slip. In fact, I reckon it 'could happen sometime' – that is, over the long term, it is 'Likely.' You'd be lucky not to get anything less than a severe cut, so that would make such an outcome 'medical attention and several days off work' or 'serious injury.' The risk numbers? 3 or 2 – and that's getting pretty dangerous.

Using jack-stands, I think the likelihood of injury from a car falling on me is 'very unlikely' – but if that did occur, the outcome would surely be 'death or serious injury.' That's dangerous indeed. I always think that working under a car on jack-stands is dangerous. I never, ever get under a car that's supported by anything less than doubled-up jack-stands – so, for example, using four stands when in fact two would support the weight. As a result of that approach, I reckon that the likelihood of injury from the car falling on

| How severely could it hurt someone? Or, how ill could it make someone? | How likely is it to occur? | | | |
| --- | --- | --- | --- | --- |
| | Very likely Could happen any time | Likely Could happen some time | Unlikely Could happen but very rarely | Very unlikely Could happen but probably never will |
| Kill or cause permanent disability or ill-health | 1 | 1 | 2 | 3 |
| Long term illness or serious injury | 1 | 2 | 3 | 4 |
| Medical attention or several days off work | 2 | 3 | 4 | 5 |
| First aid needed | 3 | 4 | 5 | 6 |

How dangerous is a metal cutting vertical bandsaw? Using the risk calculation table, an injury 'could happen sometime' – that is, over the long term, it is 'Likely.' You'd be lucky not to get anything less than a severe cut, so that would make such an outcome 'medical attention or several days off work' or 'serious injury.' The risk numbers? 3 or 2 – and that's getting dangerous. This is a machine where you should be very careful!
(Courtesy Baileigh Industrial)

me is 'Very Unlikely' – but if it did, the outcome would surely be 'death or serious injury.' That's still a '3' or '4.'

I can well remember the day, long ago when I cut my hand with a hacksaw. I was using it in a silly way, and it slipped, slicing straight into my hand. Considering my stupidity (and inexperience), I'd rate the likelihood of injury when using a hacksaw back then as 'Likely' or 'Very Likely.' That sort of cut will probably require 'Medical attention' – and in fact, I did need stitches – so the risk number works out to be '2.' A high risk – and yes, it did actually happen, so that risk rating reflects reality.

Now if you've been following, you can see that this approach takes into account training and experience of

the operator, the gear that you are wearing, the safety precautions you take, the people you have available to give you quick help … everything. I strongly recommend the idea to you.

## RISK REDUCTION

The other area that the trainer covered, and which struck me as a very effective idea, was in risk reduction. "Personal protective gear is the least effective at reducing risk," he said. That really got my attention – you mean, putting on those ear protectors or safety goggles is pointless? Well the answer to that is: in terms of reducing risk, it is in fact the least important.

The order in which you should take steps to reduce risk is this:
1. Elimination – remove the hazard completely.
2. Substitution – replace the hazardous activity with something else.
3. Isolation – minimise the chance of others coming in contact with the risk.
4. Engineering – use equipment or processes that makes it safer.
5. Administration – put in place rules and training.
6. Personal protective equipment – safety helmets, goggles, clothing, etc.

As you can see, in terms of minimising risk, personal gear is in fact the least effective! As soon as I heard this, I started thinking of my little boy (at that time, he was eight) in my home workshop. If I am using an angle grinder on steel, I give him goggles and ear protectors to wear (ie reduction of risk by use of personal protective equipment). But, according to the above list, it would be far more logical to remove him from the hazard – "Go play in the yard while I am doing this grinding." Obvious, after it's been pointed out!

And rather than just putting on goggles and a leather apron when I am heating steel red-hot to allow it to be

High quality hearing protectors are essential in all home workshops.
(Courtesy Climax)

easily bent, it might be a lot safer if I first implement an 'Engineering' step – say by clearing the floor I'll need to walk across when carrying that red-hot piece of metal in some tongs. The approach gives a structure to your thoughts, so you're less likely to miss some obvious first steps.

In summary, there are two critical ideas:

1. Evaluate the likelihood of an accident happening and how bad the results would be if it did happen, then calculate the real-world risk.
2. Remember that personal safety equipment is right down the bottom of the list in reducing risk of injury or death … not at the top, as you might believe.

So with that framework in place, what are some specifics you should be aware of?

## FIRE

Fire is a major danger in a home car workshop. Consider the mix: highly inflammable fuel and oil, machines that produce sparks that can travel far, and red-hot metal.

The first step in fire risk reduction is to ensure that flammables are stored well away from sources of ignition. In addition to fuel, paint and oils, remember that sawdust and rags are both easily set alight. Highly flammable fluids can be stored in a metal cabinet. Specific fire-proof cabinets are available for this purpose, but any steel cabinet is much better than having the fluid containers sitting on open shelves.

Most fires start very small, and so if the fire can be

At least one fire extinguisher capable of handling a wide variety of fire types should be in every home workshop.

extinguished quickly enough, little extinguishing power is needed. A simple, cheap and effective approach is to distribute fire buckets of sand around the workshop. I use four galvanised steel buckets, painted red and with 'fire' stencilled on them, for this function. One is near the welding bench, one near the belt sander, and the other two are positioned to cover the rest of the workshop. One of the buckets has been used in anger: sparks from a belt sander set a piece of cardboard smouldering.

In addition, I use two wall-mounted fire extinguishers. When selecting extinguishers, pick those that can handle a wide variety of fire types, including electrical and burning fluids. Mount them high for visibility and access, and place appropriate signs nearby. (You might not be the one who needs to use the fire buckets or extinguishers, so they need to be well labelled for people not familiar with your workshop.)

A very real potential for a damaging fire is one that occurs when you're not in the workshop. For example, a fire caused by a smouldering rag that ignites into naked flame only after you've left. To tackle this scenario, you need either automatic sprinklers (not very expensive if simple heat-sensitive sprinkler heads are used, and the plumbing is installed when the workshop is first built), and/or a fire alarm that triggers an external siren. Smoke alarms fitted with relay contacts can be used to easily build a system with an external siren.

A fire bucket filled with sand is a cheap and effective extinguisher for small fires. In addition to two fire extinguishers, I have four fire buckets like this one distributed around my workshop.

A carbon monoxide alarm. They are cheap and readily available.

## FUMES

There are two major sources of poisonous fumes in a home car workshop – cars and welding. The more immediately dangerous is the car exhaust.

Nearly everyone knows that car exhaust gases are dangerous, but few realise just how dangerous they are. As I write this, two people who live near me recently died. They were cleaning out an underground tank and had an engine-powered pump in, or near, the underground tank. The exhaust fumes killed them.

Do not run a car in an enclosed space, and be very careful in running a car inside a workshop, even with the major door open. If you are performing work where the engine will be running frequently, plumb a flexible steel tube to the car exhaust pipe and use forced air extraction to remove the exhaust gases. If you are lucky enough to have a chassis or engine dyno in your home workshop, not only must the exhaust be sealed to a free-flowing exhaust pipe that takes the fumes outside, but you also need good ventilation flow through the entire workshop space. Note that the danger from exhaust fumes increases with modified or older engines that are not using catalytic converters and may be running non-standard air/fuel ratios. Pits are dangerous in that exhaust gases can accumulate in these largely enclosed spaces.

A good move is to install a carbon monoxide alarm. These look like smoke detectors but are sensitive to the most dangerous gas in car exhaust flows. These detectors are cheap and readily available.

Fumes from welding are a more subtle poison, but the more that you investigate the toxins to which you are subject when welding, the worse it all looks. The Canadian Centre of Occupational Health and Safety states:

- Welding fluxes contain silica or fluoride and produce amorphous silica, metallic silicates and fluoride fumes.
- Fumes from mild steel welding contain mostly iron with small amounts of additive metals (chromium, nickel, manganese, molybdenum, vanadium, titanium, cobalt, copper etc).

- Stainless steels have larger amounts of chromium or nickel in the fume and lesser amounts of iron.

And then there are the coatings on the metal you are welding. These include:

- metal-working fluids, oils and rust inhibitors
- zinc on galvanised steel (vaporises to produce zinc oxide fume)
- cadmium plating
- vapours from paints and solvents
- lead oxide primer paints
- some plastic coatings

And it gets even worse! Gases produced from welding and cutting processes include:

- carbon dioxide from the decomposition of fluxes
- carbon monoxide from the breakdown of carbon dioxide shielding gas in arc welding
- ozone from the interaction of electric arc with atmospheric oxygen
- nitrogen oxides from the heating of atmospheric oxygen and nitrogen
- hydrogen chloride and phosgene (the latter used as chemical weapon in WWI) produced by the reaction between ultraviolet light and the vapours from chlorinated hydrocarbon degreasing solvents (eg trichloroethylene, TCE)

Gases are also produced from the thermal breakdown of coatings:

- polyurethane coatings can produce hydrogen cyanide, formaldehyde, carbon dioxide, carbon monoxide,

An auto-darkening welding helmet with a powered respirator. If you are doing a lot of welding, this equipment – although expensive – is a worthwhile purchase. (Courtesy Unimig)

oxides of nitrogen, and isocyanate vapours
- epoxy coatings can produce carbon dioxide and carbon monoxide
- vinyl paints can produce hydrogen chloride
- phosphate rust-inhibiting paints can release phosphine during welding processes

If that lot doesn't frighten you, I am not sure what will.

So how to minimise exposure? Remove all metal coatings prior to welding. Zinc coatings are particularly toxic, and so you should not weld galvanised steels without first removing the galvanising. If the welded metal develops a white, wispy coating near the weld, the zinc was not adequately removed. Before welding, also grind off paint and plastic coatings.

Do not weld when the surface is still wet from degreasers, cleaners, cutting oils or the like. The surface should be clean and dry.

Use forced air ventilation – passive ventilation is inadequate if welding for more than a few minutes. The air movement past the welding area should be as great as possible without removing too much heat from the workpiece (eg when oxy-acetylene welding) or blowing away the shielding gas (MIG and TIG welding). Even a cheap household fan is much better than nothing, and best of all is a purpose-designed ducted fan drawing from the welding area and exhausting to the outside air. If you intend to do a lot of welding, invest in a powered airflow welding respirator.

## HOT OR HEAVY OBJECTS

When grinding, welding, turning, milling, drilling or cutting, parts can become hot. Learn to be wary of any parts that have been subjected to these sorts of processes until you can be certain that they are not hot enough to burn. One way of achieving that is to cool them with water. Some parts that have been made very hot, eg by welding, should be allowed to cool first before being quenched (otherwise there may be metallurgical changes induced by the sudden cooling of the hot metal). However, as a final step, dunking the part in a container of water, or running a flowing tap over the part, will ensure it is cool. Near my belt linisher I keep a large stainless-steel vessel (an old cooking pot) filled with water – I dunk small parts in it to cool them.

In a car home workshop, there are plenty of heavy parts that are moved around. They might comprise the engine, gearbox – even suspension components and axles. Never be complacent when moving something that, if dropped, could injure you. Plan moves before making them, clear the area where the move is to occur (who hasn't had an engine crane stopped in its tracks when a wheel came up against a discarded cable tie?), and always use twice as many lifting straps as required. Keep all cranes, chain hoists, lifting straps and the like in good condition. Always consider what

A compressed spring has lots of potential energy, and so is dangerous. Always use quality spring compressors when swapping springs on struts. (Courtesy Toolstop)

would happen should a strap or chain slip or break – if the item falls to the ground that might be a bit sad, but if the item falls on you, you may lose a hand … or your life.

The lifting of cars is covered in much more detail in Chapter 6, but even when carefully and properly using jacks, jack-stands and the like, there are still some major dangers. Once I came across a picture that showed how a jack-stand had punched its way straight through a rusted jacking point on the car body. The top of the jack-stand ended up well inside the cabin. Especially when working on older cars undergoing restoration, remember that jack-stands depend on the integrity of the body to be safe.

Before getting under a car – whether it is supported by a hoist, jacks-stands or ramps – give the car a hearty shove sideways and ensure it doesn't move. (If using a hoist, do this with the car only just off the ground!) When lifting cars, remember that the centre of gravity is probably not at the middle point of the body, and that a car that is having its engine removed will change substantially in its distribution of mass. There are plenty of pictures around that show how cars can fall off hoists – usually because the car's centre of mass was not centred on the hoist, or the position of the centre of mass changed dramatically (eg by an engine being removed).

## OTHER CAR DANGERS

Springs in suspension systems that are subjected to preload (eg they're mounted on a MacPherson strut) must be compressed using good quality spring compressors, prior to removal or replacement. Compressed springs contain a lot of energy – if one were to escape the spring compressors, it could seriously injure or kill you. Always be very careful when working with compressed springs.

Airbag and other active restraint systems in cars are potentially very dangerous. Always read and understand the workshop manual safety instructions before dealing with these components. Yellow encased wiring is an indication that you are dealing with such a system.

Fuel tanks that require repair or modification, and which have previously been filled with fuel, are very dangerous. For example, cutting out the base of a fuel tank and welding in a new fuel sump should be undertaken only after the tank has been scrupulously cleaned of fuel and its vapours. Steam cleaning is often used to do this – I simply choose to take the tank to an expert welder in this area and leave it with them!

As hybrid and electric cars become more widespread, it is more likely that people will be working at home on these cars. The high voltage battery in these cars is very dangerous – perhaps more dangerous than most people realise. Not only does the battery have a high voltage, but it also has the ability to deliver a large amount of DC current. Again, carefully read the workshop manual and don't be at all cavalier in your approach to such systems. High voltage wiring is indicated by the use of orange insulation and covering tube.

## PROTECTIVE GEAR

While over the last few decades, professional work environments have undergone massive changes in the protective gear that is required to be worn, many people in home workshops continue to wear the minimum of gear. As described at the beginning of this chapter, wearing protective gear is not a panacea, but it will reduce the severity of injuries you might otherwise endure.

Clothing should be head-to-toe, with your arms and legs fully covered. The clothing should have no loose ends flapping, and should be made of materials that will not melt onto you in the case of a fire. Ensure that your trousers fall over your shoes (otherwise welding sparks can get into your socks!). Shoes or boots should be made of leather and contain toe protection – eg steel or Kevlar caps. Don't wear any jewellery in a home workshop – not even a wristwatch or wedding ring.

If the noise level is loud enough that voices need to be raised in order to be heard, wear ear muffs. Hearing loss is cumulative, so 'a little now and then' adds up over time. Ear muffs vary substantially in effectiveness – it's one area where buying good quality gear is worth it. Wear eye protection whenever you are using power tools – any power tools – or are using hand tools that cut. When welding, wear gloves and appropriate eye and face protection. Do not wear gloves when using power tools – gloves are too vulnerable to being caught by the tool and dragging your hand in. Note that conventional spectacles are not safety protectors, and will shatter if something large and heavy hits them.

## CHILDREN

I think it vital that children are welcome in a home workshop, and grow to love its use. But this process must be carried out in a context where safety comes first.

Firstly, children need to be aware of global dangers to which they are exposed simply by being in the home workshop. Welding flash, flying sparks, power tools resting apparently benignly on the floor. I think that this is simply a critical part of safety – they need to know what could bite, even if they are not directly involved.

They then need to be aware of the safety aspects of each tool that they use. It's easiest if they are introduced to this as they are being introduced to the tool. Always work chisels away from the body, not towards yourself. Look at where your fingers will go if the tool slips. The point of a small Phillips head screwdriver is sharp enough to penetrate

A small home workshop user. Children should be welcome in a home workshop, but appropriate safety instruction is a critical part of that process.

your skin. If the dangers are covered in the same breath as "here is how you use this tool," I think the whole spectrum of safe use is absorbed. Compare that for example with: "Now I am going to tell you about all the dangers of hand tools." How boring!

I chose to introduce my son, Alexander, from the age of about six, to hand tools only – saws, screwdrivers, spanners, a hand-powered drill, wood chisels, hammers, and files. I pointed out to him that my worst home workshop injury occurred when I was using a hand-tool in a stupid way (it was a hacksaw), but I didn't belabour the point. I monitored him from a distance, and if I saw him using a hand tool dangerously, I corrected him firmly, showing him why that approach was dangerous. When he was using a hacksaw, I made him wear safety goggles – small bits of metal can so easily damage eyes. When he was going to be spending some time in the workshop, I made him wear strong boots, and have full-length arm and leg covering.

His first 'power' tool was a hand-cranked grinding wheel (he was the power!), and he wore safety glasses when using it. At age eight, I gave him a mains-powered (corded) hand drill, a small drill press and a 3-inch variable speed mains-powered grinder. He could use those tools only when I was in the home workshop.

Taking this approach meant he could see and understand the tool progression. He understood that a 'hierarchy of danger' exists, and that with the more advanced power tools, the danger level was rising. However, having struggled with using the hand-cranked drill to go through steel, he also knew that power tools have huge advantages.

After five years of using the small bench grinder and drill press, he more often than not was preferring to use my much larger and more powerful grinder and drill press – he'd graduated to the next level. So I sold those smaller power tools. He'd also started welding with the MIG (but not the oxy), and using my wood lathe – the progression continuing. So far (and I hope that this stays the case!) he's never had any more than the most minor injury in the workshop.

So the safety aspects of having children in a home workshop are divided into two:
- The home workshop is a dangerous place for a child to be in, irrespective of what is actually going on.
- When the child is using tools, there are further, specific dangers.

If they're made aware of both aspects, there's no reason why a child cannot use a home workshop for their own creativity.

If you have compressed air in your workshop, a retractable air hose allows the hose to be extended only when necessary, reducing trip hazards. (Courtesy Toolstop)

**A purpose-designed cabinet can be used to safely store fuel, oil and other flammables. (Courtesy Toolstop)**

### ALCOHOL AND DRUGS
I have a very simple rule. If I have drunk any alcohol at all, I do not use my home workshop. If I were a recreational drug user, the same would apply. I know that some people are happy to have a beer or similar as they are working, but I strongly recommend that you do not. Any alcohol or drugs negatively impact on your coordination and judgement.

### EMERGENCY 'OFF' BUTTONS
If you have permanently installed power tools of the sort that has been covered in Chapter 3, you may want to install an emergency 'off' button on each tool. These switches are the type where if something is going wrong,

An emergency 'off' button (arrowed). These can be retrofitted very easily, and improve safety by allowing a machine to be very quickly and easily switched off if something goes wrong.

switch mechanisms, one marked 'NO' (normally open) and the other 'NC' (normally closed). If the wiring is made to the NC switch, when the button is pushed the circuit will be broken, switching off the machine. The switches I used are rated at 600V and 10A. Note that if you wish to switch off a high-current welder or similar, you may need to use a switch rated to 15A.

The switches come in a sealed box that is easily mounted on any flat surface. The mounting location is important – you want to be able to easily access the switch from any normal location you'd be standing when using the machine. That's not quite as obvious as it first looks, though – so think about it carefully before drilling the box mounting holes. For example, in the case of a table saw that I often use to cut large sheets of plywood or chipboard, I fitted two switches: one at each end of the machine. That way, both the person feeding the sheet, and the person collecting it at the other end of the saw, can access emergency buttons, as required.

The simplest way of wiring-in the switch is to insert it in the existing power cord – the machine's normal on/off switch is not altered. Here's how to do it – and note, you must be confident and knowledgeable with mains (household) wiring before proceeding.

- Switch off the power and remove the plug from the wall socket.
- Carefully strip off the outer sheath of the power cable until about 100mm (4in) of the internal wiring is accessible. Make sure you don't nick the insulation of the internal wires.
- In my Australian wiring, the earth (green/yellow) and neutral (blue) remain untouched – they simply loop through the box. The active (brown) is cut and each end is wired to the NC switch. (You must use the colour codes that apply in your country!)
- Anchor the cables going to and from the new switch so they cannot be pulled from the box.
- Screw the top of the box back into position, plug in and switch on the machine. Pressing the button should stop the machine. Always then flick the main on/off switch before turning the knob to release it.
- Taking this wiring approach makes installation quick and simple – at the cost of a reduced length of cord available to connect the machine to the wall socket.

you just slap your hand on the button and the machine immediately shuts down. The good news is that these switches are now available quite cheaply, and – if you know your way around mains power – fitting a switch to a machine can be done in just ten minutes or so.

You should buy a switch that is specifically for the function of switching off a machine in an emergency. These switches comprise a large pushbutton that, once pushed, stays locked down in the 'off' position until the knob is rotated to release it. Inside the housing are two

**Chapter 13**

# Tips and hints

In this chapter, I want to share some hints and tips, including showing you some useful workshop items that are easy to make.

## A BIG BIN (TRASH CAN)

One perennial problem in a home workshop is having a bin that's big enough to cope with all that you throw away. After years of using little bins that needed to be emptied about every other minute, I decided to go big! My local garage discards Castrol 200-litre oil drums. I took one, cut out the lid with an electric jigsaw, ground back the sharp edge, then used it as a workshop bin.

But the next step was how to get it to the trailer, so the contents could be taken to the rubbish tip? I picked up some cheap pneumatic tyred wheels, bought an axle to suit, then made a frame from square steel tube – second-grade quality, so quite cheap. For fun, I turned a handle on a new (secondhand) wood lathe I'd recently acquired. The result? A smooth-wheeling trolley that carries the bin with little effort. Looks good, works well – and the drum holds a massive amount of rubbish.

This bin and trolley combination can move a serious amount of home workshop rubbish. The drum cost nothing and the trolley was easily made.

## GREASE SPREADERS

If you're packing wheel bearings or doing another job that needs lots of grease, you'll typically use either your hands or a special grease-packing tool. But what if you want to just smear grease on a flat surface or a thread? The best thing I've found for this job is to use a tongue depressor.

Ideal for spreading grease are 'tongue depressors,' wide wooden sticks used by doctors for, yes, depressing tongues! They're available at discount shops.

Tongue what? Tongue depressors are the flat wooden strips that your doctor uses to push your tongue out of the way when he or she is looking down your throat. These sticks – available very cheaply from discount shops – are perfect for spreading grease. They are wide enough to carry a good hunk of grease, strong enough to be used as spreaders, and are cheap enough to be used once then thrown away.

## IMPROVING A METAL-CUTTING HORIZONTAL BANDSAW

Many power tools, although now very cheap, have some rough edges that can easily be improved on. A metal-cutting bandsaw that I bought was no exception.

The bandsaw has one fixed and one moving jaw in its vice. As supplied, the moving jaw tended to ride-up as it was tightened on the work-piece, leaving a small gap at the base of the moving jaw. (This is important, because the work-piece is no longer clamped securely.) To improve this, two modifications were undertaken. Firstly, a larger thick steel washer was used beneath the head of the bolt. Secondly, the rectangular nut that the jaw bolts to (and that rides on the underside of the vice slot) was shortened a little with a grinder, so that the jaw is pressed more firmly against the slide when the bolt is done up tightly.

Another shortcoming was that there was a lot of back-lash in the vice handle. This was because the whole handle and jack-screw thread could move back and forth – there was excessive end-play. It didn't change the accuracy or clamping of the vice (it still did up tightly) but it felt awful. To improve this, two washers were placed on the jack-screw, sandwiching a spring that bears against the cast base of the

This newly-purchased horizontal bandsaw was improved by some simple modifications.

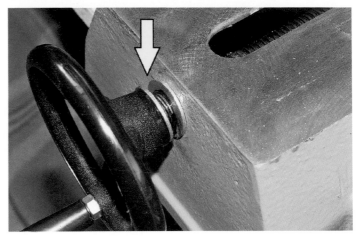

To remove backlash in the vice screw thread, two washers were placed on the jack-screw, sandwiching a spring.

saw. (You couldn't just use washers alone, as the handle and face of the base are not parallel.) This removes most of the longitudinal play in the shaft. The springs and washers are oiled to promote easy rotation.

## ADJUSTABLE HEIGHT STAND

If you cut long stock – for example, full lengths of steel tube – you'll find it hard to hold the material level and stable while it's being cut. You might have one end in the vice of the cutting machine, but without support, the other end will be

To improve the location of the moveable vice jaw, a larger thick steel washer was used beneath the head of the bolt. Secondly, the rectangular nut that the jaw bolts to (and that rides under the vice slot) was shortened so that the jaw is pressed firmly against the slide when the bolt is tightly done up.

A stand to support long stock being cut by my bandsaw or panel-saw. It uses square tube sliding within tube and a threaded knob to provide the height adjustment, while the base is an old truck rim. The stock being cut slides on slippery high-density polyurethane.

waving in the breeze. To provide support for the material, I made an adjustable height stand.

The starting point was half of a truck rim that I picked up for nothing. I welded a plate across the opening in the rim and then added a vertical post made from square tube. I then sourced some more square tube that was a sliding fit inside the upright. A nut and knob attached to a threaded bolt was used to make a lock that could hold the adjusted height. Rather than using a roller on top, I attached a piece of slippery plastic – high-density polyurethane. Using the plastic rather than a roller means the stand doesn't have to be always at exactly right-angles to the cutting machine. The stand works really well, and I also use it when cutting large sheets with my woodworking table saw.

## TUBE BEADER

If you're making your own intercooler plumbing, you're sure to have come up against an obstacle – how to form the beads on the ends of the plumbing. (A bead is the raised ridge that stops the hoses blowing off under boost.) So how do you form these beads? The answer is to make your own tool. You'll need a welder, a hydraulic press and some bits and pieces of steel.

If you look around the web, you'll find lots of people have made their own tube beader using a pair of multigrips (or big pliers), a washer and half a muffler clamp. The multigrips are used to push the washer into the inner wall of the pipe, with the outer wall supported by the bottom half of the muffler clamp. A depression is made, then the tube is rotated and another made. It's a great DIY approach – but for one problem. The physical effort is high, even for aluminium tubing. For steel, or even more so, stainless steel, the effort is too great – these materials need much more force than just hand strength.

But the fundamental approach is great – so how to give it more muscle? The answer is to use a hydraulic press – I used a 6-tonne design. The press allows you to form the bead with much less effort, and to form it in stronger materials than aluminium. However, with the power of a press behind it, the tool needs to be very stiff.

The starting point is to select a thick steel washer just

A tool to make beads like this can be made at nearly zero cost. You'll need a welder to make the tool and a hydraulic bench press to operate it.

a little smaller in diameter than the internal diameter of the pipe you'll be working with. My requirement was to work with 50mm (2in) steel tube, so the selected washer had to be a little less than 50mm (2in) in diameter. Don't use a thin 'mudguard' washer – it will bend under the forces involved. The washer I used was 4.5mm in thickness. Grind or file the sharp corner edges off the washer. If you can stiffen the thick washer by placing thinner, smaller diameter washers either side of it – great. Select a high tensile bolt and use two nuts to attach the washers to the end of a bolt. I then chose to cut off the top half of the washers.

Select some thick-wall tube that the shaft of the press will just slide into. File a depression across the end of the tube that the bolt (complete with attached washer) will fit into, then weld the bolt to the tube at right-angles. Add a gusset to stiffen the bolt against bending, while still leaving clearance for the tube to fit sufficiently over the washer that the bead will be formed a little in from the end of the tube. Make this assembly as stiff as possible. The first tool I built – just quickly to see if the approach would work – used the bolt without the gusset, but once placed in the press, the bolt quickly bent. Two nuts were welded over holes in the tube to allow the assembly to be clamped to the press's shaft – the pictures opposite show the approach.

Now for the other half of the tool. Weld the bottom part of a muffler clamp to some thick steel plate that can be clamped in the press. However, again the quick and dirty prototype showed a problem – it was hard to keep the tube square to the base and so the formed bead tended to wander around rather than being in a straight circumferential line. This problem was overcome by placing a half-round piece of slightly larger pipe in line with the clamp, so keeping the tube square when placed on the base.

So how do you use this tool? Clamp the bottom assembly to the press. Insert the top assembly on the shaft and then operate the press so that the washer lowers into the lower clamp piece. Then, with the alignment correct, tighten the bolts that hold the upper piece to the shaft. Then insert a trial piece of tube. The tube should be inserted sufficiently far that it is just touching the gusset. This sets the distance that the bead is formed from the end of the tube.

Operate the press until the washer is firmly against the inner wall of the tube and then cycle the pump a designated number of strokes. I find that on my press, two strokes of the pump form a bead of the correct height. Release the pressure, rotate the tube about one-eighth of a turn, check that the end of the tube end is just touching the gusset, then perform the same operation. Continue until the bead is fully formed.

The finished tool will make straight and even beads in minutes. I had all the bits lying around, so the out-of-pocket cost of the tool was a grand zero!

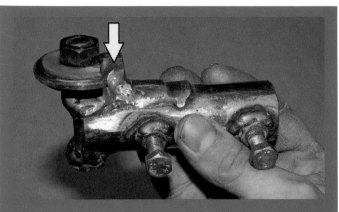

The upper part of the bead-forming tool. At left is the thick and strong washer that forms the bead depressions. The washer is mounted on a bolt, and the bolt welded across the base of a tube that, via two lock-bolts, attaches to the ram of the press. The arrow points to the gusset that prevents the bolt from bending under load.

The lower part of the bead-forming tool. It is made up of part of a U-clamp, with the arrow pointing to the curved rest that holds the tube square as the bead is being formed.

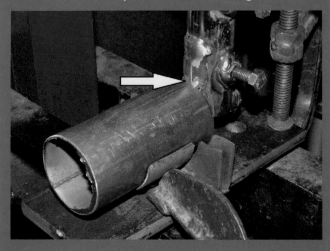

The bead being formed in the press. Form this section, turn the tube slightly and form the next section. If the tube is kept pressed up against the gusset (arrow) then the bead will be formed parallel to the end of the tube.

## SOFT VICE JAWS

Bench vices use hardened steel jaws that are embossed with serrations. These jaws very firmly hold in place items clamped in the vice, but they can also mark your workpieces. To avoid this when placing softer items in the vice, slip a pair of extra jaws over the standard vice jaws to provide protection. Soft jaws can be bought new, or you can simply cut to size two pieces of aluminium angle and slip them over the jaws as needed. I use soft jaws whenever I am holding brass, copper, aluminium and plastic items.

Soft jaws for your vice are useful in holding objects that would otherwise be marked by the standard hard steel jaws. These plastic ones are held in place by magnets.

## RESTORING EQUIPMENT

I am always on the look out for new equipment for my workshop. There are lots of items I'd really like, but, as is the case with most people, there are many pieces of desirable equipment that I simply cannot afford. Buying secondhand is one way to drop purchase costs considerably, but in many cases the newly-acquired equipment will need some restoration.

That was just the case with an old sheet metal folder that I bought. It worked okay, but looked pretty terrible with its rust and congealed grease. The first step in restoring the folder was to disassemble it. As with many old tools of this sort, it was very heavy, being formed not from welded steel fabrications but from strong castings. However, using my engine crane, I was able to completely disassemble the folder. Then it was a case of using a high-pressure water spray and detergent to de-grease, a wire brush in an angle grinder to remove old paint and rust from castings, a bench-grinder wire brush to clean up the fastenings, and plenty of effort!

I painted the main body components using a metal-specific epoxy spray. Be very careful in your paint selection as many paints designed for use on metal are quite poor, and will scratch off easily, especially on a tool that invariably will be subject to various bumps and knocks.

Reassembly of the folder was straightforward, and, as is often the case with older tools, plenty of adjustments could be made to bring tolerances back to as-new. The folder should be good, now, for another 50 years!

I bought this sheet metal folder in unloved and dirty condition. It's over 50 years old, and beautifully made of heavy-duty castings. With a bit of work and some new paint, it came up well.

I bought this laser wheel alignment machine very cheaply from a liquidator. It was mechanically in good shape but the lasers were not working.

Another item I bought and then restored is my wheel alignment machine. In a home workshop, a full wheel aligner is a bit of an extravagance (if you're on a budget, buy a magnetic bubble-type gauge instead) but when a wheel aligner appeared on a local auction site, I was intrigued. The machine, a laser design complete with cabinet and wheel turntables, was available because of a forced sale of a workshop's assets by a liquidator. It was initially said to be in good condition, however, further inquiries showed that while the lasers had previously worked, they'd now ceased to do so! As a result, the price was now even lower – in fact, it was worth buying for just the turntables and wheel clamps alone. I bought the machine and then investigated what I'd acquired.

Firstly, the mechanics were all in excellent shape – the turntables, wheel clamps and cabinet being 8/10. However, as the seller has indicated, the lasers in the alignment arms

were not working. To investigate why, I opened-up the laser heads and found some very corroded AA batteries. Even more interesting, the plastic battery holders were in parts melted! Some investigation and a little guessing indicated why. The machine was initially sold with rechargeable ni-cad batteries, and a suitable charger for these batteries was provided. However, someone in the past had replaced the batteries with conventional non-rechargeable cells. The liquidator, not knowing this, had plugged-in the charger – resulting in a battery disaster!

So what to do? The batteries had spilt their chemicals not only over the battery holders, but also over the printed circuit boards (PCBs) that were adjacent. They too were ruined. But what did the PCBs actually do? It appeared that all each did was carry a pushbutton switch and a small electronic chip that allowed the pushbutton to sequentially operate the two lasers in each head. So why not get rid of the pushbuttons and PCBs, and simply install new battery packs and two new switches in each head – one for each laser? And that's just what I did.

The machine works perfectly and, without the need to pay an hourly rate for wheel alignments, has proved to be

End-view of the restored metal folder. The unpainted metal surfaces were wire-brushed and then greased to stop them rusting.

Some investigation showed why the wheel aligner's lasers were dead – someone had tried to charge the non-rechargeable batteries! I fitted a new battery pack and two switches to allow the lasers to be individually controlled.

The laser aligner in action. The digital display shows 1.4 degrees of negative camber, and the arrow points to the laser dot that shows just over 1mm of toe-in.

A spanner (wrench) I made to allow me to easily move the toe-adjusters on one of my cars, while doing wheel alignments. (It was a cheap supermarket spanner to start with!) When you need a special tool, it's often easiest to make it.

able to tell some very interesting tales. For example, it's easy to measure bump-steer variations at different ride heights – but that's another story.

Always be on the look out for secondhand garage and workshop equipment – if it's cheap enough and was originally a quality item, it's often worth buying.

## MAKING YOUR OWN TOOLS

A good tool set will let you carry out most jobs that you need to perform, but occasionally something pops up that requires a special tool. In that case, it's often easiest to make one, especially if it needs to do the job once or twice only.

The first special tool I made was to allow me to adjust idle mixtures on an L-Jetronic BMW – my third car. The adjusting screw was under the vane airflow meter, and all the screwdrivers I had were too long and large in handle diameter to allow access. I bought a cheap 'stubby' screwdriver with a plastic handle, and then used a hammer to shatter the handle, allowing it to be easily removed. Some electrical tape wound around the metal stub gave enough purchase for my fingers to turn it – and, voila, I had my 'BMW special tool.'

Another tool I made for working on the BMW was a valve spring depressor. When I rebuilt the engine, I found that all normal valve spring compressors didn't fit – the valve springs were too deeply recessed in the head. I got a chunk of steel, and using a drill, hacksaw and file, made it into a U-shaped adaptor about as big as a matchbox. It fitted under the arm of the spring depressor and worked perfectly.

When I was rebuilding an automatic transmission on another car, I found that I needed a torque wrench that could accurately measure very small torques. I investigated buying one and discovered, to my dismay, that they were very expensive. To get around the requirement of buying a specialised torque wrench for just one or two uses, I used a normal ratchet handle and a spring balance. The spring balance pulled at right-angles to the ratchet handle, doing this at a precisely measured distance from the socket axis. To prevent the spring balance hook from sliding along the

handle, things were lined up, I wrapped around the handles two lots of electrical tape, spaced just far enough apart for the hook to pull on the ratchet handle at the right point. By reading off the value on the spring balance, this approach allowed me to torque the internal fasteners to the correct value.

And another torque wrench example – but this time at the other end of the scale. Sam Sharp, a US enthusiast who reviewed each of these chapters, comments: "A friend had the front-end apart on a large motor home. The spindle nut called for a huge amount of torque, something like 500ft-lb. We didn't have a torque wrench that would go that high, so we just used a piece of pipe to extend a big breaker bar, weighed one of us, calculated how far out on the pipe we needed to be, and sat on the pipe at the right place. Instant correct torque!"

More recently, when doing wheel alignments on my Honda Insight, I found that it was awkward to turn the toe adjusters. You almost needed a spanner (wrench) that was long and then had a tight right-angle turn at the end. I don't think that any such spanner is available – but it's easy to make one. I selected a cheap supermarket-type spanner from my collection and cut off the open-ended part. I then welded this to a bar at right-angles, and then welded another bar across the end, making a T-shaped tool with the spanner claw at the end. Using the same approach, I think I'll make another two similar tools (in different sizes) so that I can also tighten the locking nuts on the tie-rods. That's why it's good to pick up cheap spanners when you see them!

If you do a lot of work on your car, you need some of this! It's a tub of anti-seize compound – it can be used on any nuts or bolts that you'd like to be able to easily undo at a later stage.

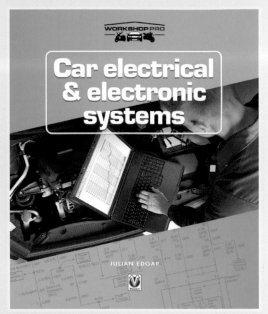
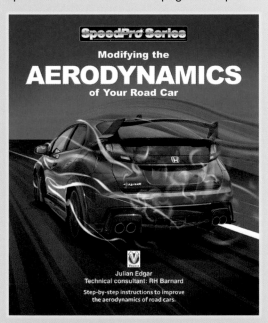
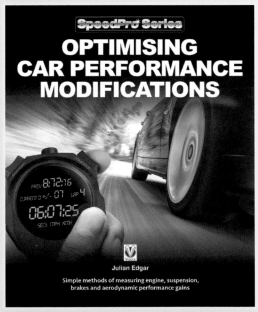

One of Veloce's *Classic Reprints*, this full colour book provides clear and complete information for the classic enthusiast who wishes to service, repair or improve car electrical systems.

ISBN: 978-1-787111-01-1
Paperback • 27x20.7cm
96 pages • 301 colour pictures

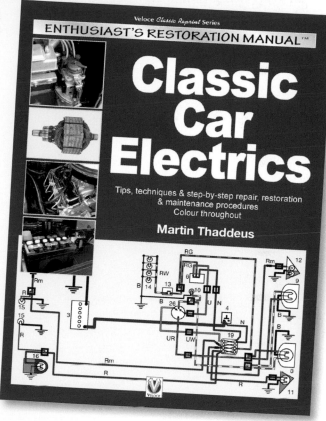

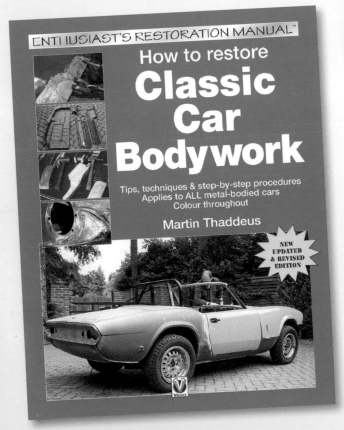

This book was written for the home restorer who, until now, may have lacked the confidence to tackle bodywork. With specially devised techniques that don't rely on workshop plant, this work bridges the gap between professional and amateur. The text is readable, the photos bright, and instruction clear. A real boon for the classic car enthusiast.

ISBN: 978-1-787111-67-7
Paperback • 27x20.7cm
128 pages • 350 colour pictures

email: info@veloce.co.uk • Tel: +44(0)1305 260068

# INDEX